AF808974

THE INTERLEUKIN REVOLUTION

Also by
Kendall Smith

Interleukin-2

The Quantal Theory of Immunity:
The Molecular Basis of Autoimmunity and Leukemia

Molecular Immunity: A Chronology of 60 Years of Discovery

THE INTERLEUKIN REVOLUTION

Kendall Arthur Smith

The Interleukin Revolution
Copyright © 2023 by Kendall Arthur Smith

This is a memoir. It reflects the author's present recollections of experiences over a period of years. Certain names, locations, and identifying characteristics have been changed, and certain individuals are composites. Dialogue and events have been recreated from memory and, in some cases, have been compressed to convey the substance of what was said or what occurred.

All rights reserved. No part of this book may be used or reproduced in any form, electronic or mechanical, including photocopying, recording, or scanning into any information storage and retrieval system, without written permission from the author except in the case of brief quotation embodied in critical articles and reviews.

Book design by The Troy Book Makers
Printed in the United States of America
The Troy Book Makers • Troy, New York • thetroybookmakers.com

ISBN: 978-1-61468-798-6

For Lynn, who took me the rest of the way

CONTENTS

CONTENTS

PREFACE

WITH THE PRESENT PANDEMIC OF Covid-19 we have all seen in living color and real time a scrouge sweep the world and be met in battle by the immune system. This book was written during the pandemic as everyone was trying to grasp the fast-moving public health policies and understand the personal health dangers, while hoping for a vaccine. Immunity was suddenly a topic of the daily news.

Immunity is impressive. But how does it work? Progress in our knowledge has been a long time coming. Since our present understanding of the immune system is so new, most people have no idea of how their bodies respond to invasion by dangerous microbes.

Throughout history plagues came and went, often killing major portions of the world's population. It was not until 1798 that Edward Jenner, an English country doctor reported the development of the first vaccine, against smallpox, a devastating disease that accounted for ten percent mortality around the globe. Jenner showed that cowpox, a benign affliction, protected against the much more severe disease of smallpox.

Almost a century went by until the 1880s when Louis Pasteur popularized the notion that all contagious maladies might be prevented by vaccination. However, without an understanding of immunity, vaccines proved difficult to develop. In 1918, now a century ago, the first influenza pandemic of the modern era killed 50 million of the world's 2 billion people.

It was obvious that immunity had to be understood.

Throughout the first half of the 20th century antibodies were found to be molecules in the blood important for immunity. By 1950 antibodies had been traced to originate in white blood cells and by 1960 the structures of antibody molecules became elucidated.

Then, around 1970 it was established that there are at least two kinds of white blood cells involved in immunity, B cells and T cells. B cells were found to produce antibodies, and T cells proved to reject organ transplants and to combat infections, especially viral infections, like smallpox, influenza, and Covid-19.

Then the paramount question became how do T cells function?

T cells do not produce antibodies, but they do make molecules called interleukins. This is the story of an ongoing revolution in science and health care ushered in by my discovery of the first interleukin molecule in the early 1980s. Now 40 years later, more than 40 such molecules have been found to be hormones of the immune system. We found that interleukins orchestrate and regulate immune responses to microbes through interaction with receptor molecules, much as do hormones. This realization has opened the way to the many new immunotherapies that we enjoy today for treating autoimmunity, allergy, infections and even cancer.

Of utmost importance, as a result of the new comprehension of how interleukins regulate immune responses, we are in a much better position to understand the Covid-19 pandemic, and the immunity generated by the coronavirus vaccines, which have allowed societies to recover.

This book takes you into the inside world of my research laboratory to relate how we began to comprehend the nature and importance of interleukins. These stories illustrate the excitement of the quest and the euphoria of discovery that drives all scientists to uncover new knowledge.

Enjoy the journey.

PROLOGUE

ORDINARILY THE PROFESSOR'S SERVICE AT Yale New Haven Hospital was a benign rotation, a time to recover from chronic sleep deprivation. No one was terribly sick since most patients were in the house for diagnostic workups. However, the year was 1968, and the Hong Kong flu pandemic was in full force. I was a medical intern wrapping up my chart reviews when I got a call from the emergency room. They were sending me a very sick patient on a ventilator. I cleaned up my scutwork and waited. Around midnight, two hours after I'd received the alert, a team of nurses and doctors wheeled a patient on a hospital bed off the elevator and delivered him to one of the only remaining vacant rooms in the hospital. Someone handed me the lab data obtained in the emergency room, which revealed a very low blood oxygen saturation prior to placement of an endotracheal tube.

My new patient was comatose, having been heavily sedated to allow the ventilator to take over from his normal breathing reflex. I ordered a bedside portable chest X-ray. It showed both lungs opaque to the X-rays. They looked like the lungs of someone who had drowned, which is essentially what was happening to the man. All through the night, as the ventilator cycled on and off, I could barely hear breath sounds coming from his lungs' tiny air sacs. They had filled up with fluid. Repeat blood-oxygen determinations showed a progressive deterioration. In the wee hours of the morning, my patient underwent a cardiac arrest. Although we worked for hours, it was impossible to resuscitate him.

As it happened, this patient was a special one. He was one of Yale's first kidney transplant recipients. He'd been given heavy doses of drugs to prevent his immune system from rejecting the kidney.

The term *immune* is from Latin, meaning exempt. With his immunity, or exemption from infections suppressed, this therapy left him susceptible to microscopic organisms. Ordinarily innocuous, these bugs — bacteria, viruses, even single-celled parasites — become opportunists, invading an immune-crippled host. Bacteria can be seen with an ordinary light microscope (magnification 1,000X) while viruses are much smaller so an electron microscope is needed (magnification 1,000,000X). And because the drugs suppressed the immune system, which normally begins rejecting the invasion of microbes soon after infection, the patient was a sitting duck. At the time, we knew so little about how the immune system functions and we had no antiviral drugs effective against flu virus. We are still woefully unprepared to combat most viral infections, as evidenced by the Covid-19 pandemic. Back then, without drugs to kill the virus and a complete lack of understanding about how the immune system functions, there was nothing more we could do therapeutically to reverse the man's inevitable decline and death.

Now, more than 50 years later, we know that immunosuppressive drugs impair the *innate*, or inborn, immune response that early on normally impedes the replication of viral invaders. Thus, unfettered for several days after infection, there is an abnormally swift multiplication of the viral particles so that days later, when the *adaptive*, or acquired, immune system finally kicks in, the result is a tremendous battle between the expanded number of microbes and the struggling immune response. It was the collateral damage from this second, immune system response that killed my patient by what we now call immunopathology. A similar immunopathology is responsible for the severe illness and death brought on by SARS-CoV-2, the coronavirus that causes Covid-19.

During the 1970s and 1980s, my laboratory made a series of discoveries at Dartmouth Medical School that revealed what we know today. The immune system is regulated by its own unique hormonal molecules, called *interleukins*. We now understand how interleukins respond to infections. And more importantly, our grasp of the hormonal regulation of immunity makes it possible to manipulate the immune response therapeutically, either by suppressing or augmenting these hormones, their receptors or their capacities to signal the cells of the immune system.

This book tells the stories of the discoveries of the interleukin immunological hormone molecules that made for a revolution in immunology. This story began 50 years ago, is still going on, but remains largely unknown by most people. To achieve a level of comprehension of any new science, it is necessary to follow the chronology of the acquisition of knowledge. This historical approach to learning science in this memoir follows the questions that consumed me and my team, how we formulated our hypotheses, how we developed novel experiments to monitor and test for newly discovered molecules, how we interpreted our results and how our lives were changed in the process.

Our progress opened the way to unprecedented new therapies.

CHAPTER 1

Learning the Language of Medicine

I HAD BEEN ON A straight track to medicine ever since junior high school when I decided I wanted to become a doctor. At that time, in the mid-1950s, physicians were atop the hierarchy of people respected in the community. My revelation came upon me one summer as I caddied at the country club near my neighborhood in Fairlawn, Ohio, a suburb of Akron. On the weekends, the doctors played golf. They all seemed to drive Cadillacs, and to top it off, they had the best-looking wives. I decided I wanted to be just like these men — a rich American doctor.

I studied hard in high school, went off to a small liberal arts college, Denison University, majored in biology, then entered the Ohio State University College of Medicine. From the first day at Ohio State, we all wore a white jacket and necktie, and I felt like a real doctor. I loved medical school. It seemed like every day I learned something new. Class was from 8:00 a.m. to 5:00 p.m. in amphitheater-like lecture halls. After an hour break for dinner, I would study at the kitchen table, usually until midnight. I roomed with a friend from high school who was in dental school. We usually took Friday night off. In the morning we were back at our books, studying all day Saturday and Sunday. Our one-bedroom apartment in a new building overlooked a parking lot not far from the campus. It was not much of a view, but that made little difference because we were at school all day and at night it was too dark to see much anyway.

It was fun that first year to learn the way the body works and the names, usually in Latin, for every bone, muscle, nerve, artery and vein. Thankfully, I honed the skill to memorize in my college organic chemistry course so that by the time I got to med school, I had mastered it. Just reading something silently or even out loud was not sufficient. I could remember new information only when I wrote it down, over and over. In retrospect, med school was really a language school for learning, then speaking a whole new language. That first year they threw a prodigious amount of information at us. The good news was I learned that hard work really does pay off when, at the end of the year, I was pleasantly surprised to receive the Most Outstanding First-Year Student award.

It was 1965. That summer, I married my high school sweetheart. Lynn and I started dating in 1959 when she was 16 and I was 17. On

Heading off to the reception, July 3, 1965.

our wedding day, we both felt as if we would never be more mature. Lynn was truly gorgeous (still is), and I had achieved one of my original goals, to have a good-looking wife. At the reception after the ceremony, my mother took Lynn aside and told her, "We got him this far, sweetie. Now it's up to you to take him the rest of the way." Lynn told me about this many years later. I never asked Mom what she meant, but Lynn says that now she knows.

The med school curriculum was divided between studies of the normal human body the first year, followed in the second year by the abnormal or diseased body. Microbiology, the study of pathogenic (disease-causing) microbes, was one of the most important courses. This is where we were introduced to the immune system. It was truly just an introduction because we spent only a couple of days on the subject. One would think that something as important as the immune system deserved its own course. I didn't know at the time, but it was one of the bodily systems about which little was understood even though Jenner and Pasteur had introduced immunology via vaccination more than a century earlier. By the early 20th century, severe allergic reactions were encountered after repetitive injections of vaccines, signaling that immunity was more complicated than it was originally thought to be. As a result, the study of immunity became the focus of more fundamental research by scientists, and immunology became a basic science, the purview of Ph.D.s rather than M.D.s.

Early in my first clinical rotation on infectious diseases in the third year of med school, as I walked across the medicine quad one day, I ran into a friend in his fourth year and in the throes of applying for his internship, the first postgraduate year of M.D. training. Right off, he asked me where I was going to apply for my internship. At this stage, I had no idea what kind of a doctor I wanted to be, let alone where to apply for that year of training. I told him that, adding, "but I know where I don't want to go, and that's Vietnam." To which he immediately replied, "You should talk to Chuck Mengel. He is the only one here at OSU who can get you into the NIH."

"What's the NIH?" I asked. My friend educated me that it stood for the National Institutes of Health. Of utmost importance, he told me that a two-year stint at the NIH would satisfy my military service requirement because I would be in the U.S. Public Health Service.

He advised me, "When you talk to Mengel, don't tell him you don't know what kind of a doctor you want to be. Tell him that you want to go into academic medicine, that you want to get a good internship, and that you want to go to the NIH." That's exactly what I did.

Dr. Mengel was new on the faculty, having been recruited just a year or so earlier from Duke to be the chief of the Division of Hematology and Oncology (blood and cancer) at Ohio State. Tall and charismatic, about 40 years old, he arrived with the reputation of being a terror with students and the young doctors on his rounds. He conducted them by the Socratic method, challenging his large following with questions to which he knew the answers and was pretty sure the others did not. The day I sat in his office and gave him my prepared talk about wanting to do academic medicine and go to the NIH, he said, "OK, Smith" — he never called me by my first name — "you'll have to come and work in my lab your entire senior year." This idea shocked me. I immediately voiced my concern that I would miss all kinds of clinical training in the hospital. "Not to worry," he said, "it'll be a little rough the first six weeks, but you'll get the hang of it and then it'll be a piece of cake."

Midway through that third year during my Medicine rotation, I was sent down to the ER to admit a patient to our ward. I was introduced to a frail, middle-aged man in acute distress. He was breathing at such a rapid rate that he could barely talk to me. His initial blood tests revealed severe anemia, which probably explained his rapid breathing: he was not receiving enough oxygen via his deficiency of red blood cells (RBCs). I rapidly took him up to the Medicine floor and called an emergency hematology consultation. The hematologist inserted a large needle into the pelvic bone to aspirate a sample to examine the patient's blood-forming cells. I ordered myriad other tests and gave the man oxygen and multiple blood transfusions. It turned out that he was an alcoholic who, after repeated bouts of acute pancreatitis, was suffering from chronic pancreatic insufficiency. The pancreas is a gland in the abdomen that secretes digestive enzymes into the small bowel that normally break down our ingested food into molecules small enough to be absorbed into the bloodstream. We kept him in the hospital for three months in the Clinical Research Unit and conducted a series of tests that revealed his pancreatic insufficiency led to a malabsorption of vitamin B12, which is required to form new blood cells. Ultimately the saga uncovered a new cause of pernicious anemia, and resulted in my first paper, published as a case report in the *Archives of Internal Medicine* (1).

This experience introduced me to Dr. Norton Greenberger, chief of the Division of Gastroenterology, who co-authored the report along with my senior resident and who had just joined the faculty, coming from Massachusetts General Hospital in Boston. One day in his lab while discussing my case, he asked me what branch of medicine I was going into. Still an unformed piece of clay, I asked him what he thought were the best specialties. Without hesitation, he said immunology. "It is the coming new field where very important questions are going to be answered in the next decades." I had not expected this coming from a GI doc, and I did not forget his advice.

Part of my time during my fourth year I spent interviewing for internships. Lynn and I took an eastern trip, first to New Haven, Conn., to Yale, then down to Baltimore to Johns Hopkins Hospital, and on to Durham, N.C., to Duke. I felt at home at Yale, where they seemed pleased that I had applied. Hopkins not so much. The residents at Hopkins took me to the rounds of the chairman of Medicine, A. McGee Harvey, who was a legend and was treated like a god by the young doctors. However, Baltimore was clearly a segregated city. The medical center, which was in a particularly depressed section of town, was surrounded by a chain-link fence topped with barbed wire and patrolled by police armed with automatic rifles and partnered with German shepherds. That's all that Lynn needed to rule out Hopkins. Duke University Medical Center was our first exposure to the Deep South. There we found evidence of the Jim Crow laws foreboding. But we were elated in March when the intern match placed me at Yale New Haven Hospital. Soon thereafter I heard from the NIH that I had been accepted to become a clinical associate at the National Cancer Institute, two years hence, after my internship and first year of residency in internal medicine at Yale.

The end of my senior year of med school in 1968 culminated in a banquet and awards ceremony where I learned that I was to graduate summa cum laude, the highest honor. Only three students from our class of 150 were chosen for this distinction. Even more marvelous, one of my best friends, Jeffrey Wardwell, was also selected. Jeffrey and I had been friends (and competitors) since high school. We went to the same college, roomed together, pledged the same fraternity (Phi Gamma Delta), and Jeffrey married Lynn's college roommate, Marjorie Culloden. We're still best friends today and we visit each other at least once a year. The rank order of the three best students, first, second or third, was not revealed, but Jeffrey and I each assumed who was number one. Obviously, I had to be number one, while Jeffery thought he was. Anyway, it was good to be honored. Talk about a mind-rush. I realized what euphoria was. It was a fitting end to my childhood dream of becoming a doctor. Of course, Lynn was very proud of me.

As I progressed through the ranks to internship and residency, I began to realize that no one knew the causes of most of the major diseases that brought patients to the hospital. Moreover, nearly all the therapies available were not curative. They only slowed the progression of the diseases to an inevitable demise. It seemed to me that there had to be a better way.

The first day of my internship in June coincided with a command performance to appear at the NIH for an interview to de-

Geoffrey Wardwell and me just after the graduation ceremony, 1968.

termine which program to which I would be assigned. Instead of reporting for duty at 8:00 a.m. at Yale New Haven Hospital, I flew to D.C. As the plane circled over the nation's capital to land at Washington (now Reagan) National Airport, I had a bird's-eye view of the National Mall, which at the time was the site of Resurrection City. It was part of the Poor People's Campaign envisioned by Martin Luther King Jr. though he never made it there. King was assassinated in April. I caught a taxi to take me to the NIH in Bethesda, Md., and as we drove through the deserted streets of the capital, half-tracks with armed soldiers were the only other traffic. It was surreal.

After a full day of interviews with various lab chiefs, I hurriedly flew back to New Haven and reported for duty at 6:30 p.m. eager to get started in my starched whites. The first-year resident in charge of our floor, who would be moving on from Yale the next day, greeted me, "Well, if it isn't finally Dr. Smith, the missing intern. Welcome. We have your first patient ready and waiting for you." Whereupon he ushered me into a small treatment room next to the nurses' station (the ICU was full) where a young woman lay comatose on a respirator. Though I could not take a history, I did a physical exam. I could find no glaring pathologies and had no idea what was wrong with her. You can imagine at this point how scared I was. On the verge of a panic attack, I went back to the resident, who belittled me in front of several other interns for what seemed like an eternity before he finally told me that my patient had imbibed a sedative overdose. That night, I never went to bed but stayed up monitoring her vital signs, intake of IV fluids and urine output every hour. The idea was to wash out the drug. I watched from the treatment room as the sun slowly rose over the city and said to myself, "This is insane. I cannot do this for a whole year."

Dr. Mengel was correct. The first six weeks were rough, but I gradually got the hang of it and started speaking the language I had learned in med school. I stayed sleep-deprived the entire year since interns were on call every other night and weekend, which meant we practically never slept. This tradition was de rigueur, established and maintained by hospitals to care for their patients for the least cost. Young doctors were called interns the first year because the hospitals only supplied room and board plus a small stipend. By tradition most new docs who were unmarried lived in the hospital, so that after the first year they were called residents. It was fortuitous that the year just prior to my internship, a court ruled that interns and residents were not students, as claimed by the hospitals, but rather hospital employees, subject to income tax. My projected $3,000 stipend as a "student" was doubled to a salary of $6,000 a year as an employee. This was the same as Lynn's salary as a first-grade schoolteacher back in Columbus, Ohio.

We could now get by on my paycheck and Lynn no longer had to work to support us. This was fortunate because Lynn had become pregnant. She was so excited to be able to live her dream of motherhood, the attainment of the storybook life portrayed in the movies and on TV at the time. Think *Father Knows Best* and *The Adventures of Ozzie and Harriet*. We found a great townhouse apartment in a new complex that fronted on the Long Island Sound in West Haven. Lynn took long walks on the beach with Slick, our rambunctious foxhound and surrogate child, while I toiled away in the hospital.

In October, Lynn went into labor several weeks prematurely. I had been on call all weekend and came home after a full day on Monday evening. We had just gone to bed when Lynn announced suddenly that her water had broken. I couldn't believe it. I was so tired. My first thought was maybe that she had just wet the bed. Lynn adamantly nixed that idea. I rushed to get my textbook to read up on what we should do. Of course, Lynn knew what we should do. We packed and headed off to the hospital. We were both excited. The adrenalin was pumping. However, after three days of drug-induced labor, the baby had not moved in the birth canal. Lynn was exhausted and so was I. Finally, she spiked a fever, which was a sign of amnionitis (an infection in the womb). That required an emergency C-section. I wasn't allowed in the delivery room. It was a very good thing because when the ob-gyn docs opened the uterus, they found the amniotic fluid thick with pus. This really shook up the docs and they immediately jerked the baby out. In the process, the infected amniotic fluid spilled into Lynn's peritoneal cavity. Rather than start broad-spectrum antibiotics right away, they put her in isolation in a private room where I volunteered to stay with her because they were short-staffed of nurses. The isolation was necessary because a few months earlier Lynn had contracted food poisoning. Now she was carrying, or colonized, by salmonella, the bacterium that causes severe gastroenteritis. The docs were scared to death that she might infect their whole ward with this bug, which was notoriously difficult to eradicate with antibiotics.

Around two in the morning, Lynn spiked a fever to 103° and had a shaking chill. I immediately alerted the nurses, who called the ob-gyn resident. He prescribed an antibiotic over the phone but did not come to see her, nor did he draw any blood cultures. I was beside myself. I knew this was wrong. It had been drummed into me that one never started antibiotics without first taking samples from the patient to culture in the lab in hopes of identifying the bacteria responsible for the fever. I went up to the Medicine floor to get the blood-culture kit and drew the samples myself. This break in hospital protocol of a medical intern caring for his wife on an ob-gyn floor created a brouhaha. The next day the chair of Medicine came to see Lynn and told me to relinquish my volunteer nursing duties. I had breached protocol in a tradition-bound system, even though what I did was medically correct. I was dispatched back to the Medicine floor.

It proved to be good that I had taken the blood cultures because they were positive for three different bacteria, one of which was salmonella. The sole antibiotic originally prescribed by the resident would not have covered them all. The same three bugs were eventually cultured from the amniotic fluid. The fact that the blood cultures were positive proved that the bacteria had invaded the blood system from the abdominal cavity and that Lynn was suffering from what is termed *sepsis* (from Greek, meaning rotten) caused by microbes invading the blood. The common phrase is blood poisoning. The next morning the Infectious Disease Division was consulted to direct the antibiotic therapy. Also, the chair of Medicine sent Lynn a huge bouquet of flowers, came to visit her and apologized to me for his original reaction to my unorthodox behavior. Lynn could have died. Even now, 1.7 million Americans become septic each year and 270,000 die, according to the Centers for Disease Control. The only hope is to stop the replication of the bugs with antibiotics before the immune system is overwhelmed by the invading army. The fact that the bugs had gained entrance to the blood was evidence that the immune system had already been breached. In our case, the blood cultures saved the day because the proper antibiotics were eventually prescribed and killed all three microbes, thereby giving the immune system time to respond and to mop up after the microbial slaughter.

I was exonerated and allowed to finish my internship and subsequent year of residency, which I had not even realized was in jeopardy. Lynn and our new son Kerrington — soon to be known to all as Kerry — did not leave the hospital for two weeks. Lynn lost about 20 pounds from her pre-pregnancy weight during the stay. She weighed only 85 pounds when she came home. Kerry had spent the time in the pediatric ICU and, despite all fears, never became infected. The pediatric docs started him on antibiotics right after the delivery and adjusted them according to the results of the blood cultures I had taken. Several weeks later, Lynn's case was presented at ob-gyn grand rounds as a lesson in what *not* to do.

When I started my internship, I worried because I had completed med school without ever seeing patients suffering from a heart attack or many other common maladies. Since 25 percent of American deaths every year were the result of heart attacks, I knew I was ill-prepared. Not to worry. I soon saw many patients with heart attacks and, as Dr. Mengel promised, I became an old hand at managing them. Early one morning in the ER, a gurney was wheeled in from an ambulance with a young man (38 years old) who had had suffered a massive heart attack. No sooner than we got him into a room, he went into cardiac arrest. Cardiac resuscitation is truly a sight to behold. Usually, utter chaos ensues with a dozen docs and nurses working on the patient. One is giving chest compression, another inserting a tube through the mouth into the breathing tube, some hooking up an EKG machine, others getting an IV started. We tried to resuscitate the man to restart his heart with a defibrillator to no avail. After several hours we had to call it quits. It was my responsibility to inform the family. When I entered the waiting room, I found a young woman with her two daughters who could not have been more than 8 to 10 years old. It was not easy to break the news. I could not believe that such a young man could die so suddenly. As I learned more, I realized no one really knew why heart disease occurred. During the 1950s, for the first time it was documented on autopsies that many young American soldiers killed in the Korean War already had severe atherosclerosis (a disease of the arteries with inflammation and scarring of the inner walls, leading to heart attacks and strokes). Subsequently, it was found that even infants had evidence of atherosclerosis. No one knew why the inflammation had occurred and why the disease remained asymptomatic until a heart attack made itself known to both patient and doctor years later. By then, though, it was too late to reverse a lifelong disease course.

Another time, also in the ER, a fellow in his 60s was brought in at 6:00 a.m. He was hyperventilating, in a lot of pain, and was cold and clammy, but his vital signs were stable. I gave him morphine and oxygen and had him sitting upright on the gurney so he could breathe more easily. As his pain diminished, he became less agitated. Just before he was taken upstairs, he called me over and whispered "Doc, this all happened when we were making love. Does this mean that I will never be able to again?" I tried to reassure him as best I could and told him his cardiologist would be able to counsel him on that and all other physical activities when he was discharged from the hospital. I did not really know how to respond to him. But I knew that his life was going to change drastically, that he would be a "heart patient," that this meant he would be placed on a low-salt, low-fat, high-carbohydrate diet, which many find essentially tasteless, and that he would be told to exercise minimally to avoid placing a strain on his heart. In essence, he would become a "cardiac cripple," an example of an iatrogenic (that is, physician-induced) affliction of modern medicine, and he would be afraid to move strenuously, let alone engage in sex.

We still do not know the cause of blood-vessel disease. Only now are we beginning to understand that a high-carbohydrate, low-fat diet championed by the cardiologists promotes atherosclerosis through some unknown mechanism that initiates and sustains inflammation of the arteries. Eventually this results in fat deposition and scar formation in the arterial walls throughout the body. We now are aware there is a mysterious connection between our American diet and the immune system, still yet to be defined.

Lynn's ob-gyn doc pleaded with her not to have another baby at Yale New Haven, so as not to risk repeating the horror show, but she turned up pregnant again (oops) six months after Kerry's birth. This time the docs were skittish — surprise, surprise — about letting Lynn go into labor because of her history, so she had a second C-section. I waited in the surgical suite while outside the window from my second-floor perch I followed demonstrations in support of the Black Panthers. They were about to be tried on charges of murdering a fellow member of their radical organization whom they suspected of being a government informant. All the stores fronting the New Haven Green were boarded up. Armed military units patrolled the deserted streets. At the hospital, all went well this time. Our second son Kelden Michael — Kelly, we called him right away — was born on April 28, 1970. Mother and baby were both very happy, as was I. The trial began a few days later without any major disruptions.

All during these years the Vietnam War was ramping up. Trained docs were needed to take care of the sick and wounded. Under a program known as the Berry Plan physicians were allowed to defer military service until they completed their specialty training, so that most of my classmates avoided the draft via the Berry Plan. As fate would have it, the NIH sent me to Baltimore in 1970 to join a branch of the National Cancer Institute, which was a far sight better than Da Nang. Even though Lynn and I had nixed Baltimore for intern and residency training, we ended up there anyway.

Throughout the next year as one of 11 clinical associates, I cared for patients who were volunteer subjects in experimental chemotherapy trials. I had been trained in internal medicine, but they needed more docs for the neurosurgery service, which was investigating new chemotherapeutic drugs for brain tumors. I was not happy with this assignment, being an internist, but as I was "in the army now," and absolutely nothing could be done about it. At least I did not have to scrub for surgery. Instead, I functioned on the floor to care for the patients pre- and post-op. I did learn how to give a thorough neurological exam. I also cared for children afflicted with the most common type of childhood leukemia in which I had a much greater interest because at least it was medicine and not surgery.

At first, I relished this new post. I was on call at the hospital only every eleventh night instead of every other night as an intern or every third night as a resident. Lynn was also quite pleased, to be sure, since I would be home more and especially because my salary doubled in this transition. That meant that we left chuck steaks behind for porterhouse from the PX at the nearby Aberdeen Proving

Ground, which would become infamous for testing Agent Orange during the Vietnam War. Thanks to the government subsidies, everything was half price at the PX. After several months, I found that I could also double my salary by moonlighting one night a week at Bon Secours Hospital, which was in a rough section of West Baltimore. I covered the ER from 7:00 p.m. until 7 the next morning. In essence, this meant that I didn't sleep that night. Patients came through the doors nonstop. I saw more gunshot wounds and stabbings there than I probably would have ever seen in Vietnam. However, for the first time in our lives, it felt as if we were flush with money.

Midway along my first year at the Baltimore Cancer Research Center, I was terribly upset when one of my patients died. The 12-year-old girl had suffered from leukemia for several years. She had undergone multiple rounds of combination chemotherapy, which included high doses of prednisone (a synthetic steroid hormone called a glucocorticoid because it regulates the metabolism of glucose) and several cytotoxic (cell killer) drugs. These drugs had been found to yield dramatic but transient kill-offs of leukemia cells. But the side effects were devastating because they also were toxic to most normal cells in the body. In particular, glucocorticoids alter how the body metabolizes, or burns, the three primary nutrients: proteins, fats and carbohydrates for energy. These three kinds of molecules are essential for life. Proteins are comprised of long chains of amino acids and function primarily as the tissue structures of the body, like muscles. Fats are comprised of cholesterol and long chains of molecules called fatty acids. Fats function primarily as energy storage molecules and as key components of all cell membranes. Carbohydrates include sugars and starches, which are polymers, a large number of individual molecules bonded together. Polymer is from Greek, *polumeros*, having many parts. Carbohydrates function as a source of energy for our bodies and also as structural components of proteins and fats. Under the influence of glucocorticoids, the proteins in muscles and other structures are broken down and used as fuel. As a result, the extremities can become wasted. Meanwhile fat is not used for fuel and accumulates in the face, trunk and abdomen (so-called moon faces and buffalo hump). Sugars and starches are also not used for energy efficiently, causing patients to develop diabetes. The once-pretty, young girl was unrecognizable.

The immune system is suppressed by leukemia as well as by steroids and chemotherapy. This makes it more difficult to fight infections, just like my patient with the Hong Kong flu. Immunosuppression had created a situation where the girl's body was left defenseless against the many microbes in the environment, which ultimately led to her death due to sepsis. I was devastated after having been immersed in her care for several months. I left her bedside when it was clear that she was gone. Despite all the clinical training I'd received, this was the first situation where I was intimately involved with a patient and her family. When I had to tell the girl's mother of her death, I found it difficult to suppress my tears. The mother, whose life and family had been disrupted and consumed by years of focusing on this daughter to the exclusion of her other children, appeared relieved that it was finally all over. She told me she hoped my tears were legitimate. I realized she was cynical about the whole process with us trying to keep her daughter alive at all costs. I was deeply depressed, wondering what I was doing by providing enormously toxic chemotherapy. I told Lynn, "The patients are extremely sick when they come to us. Then, when we administer our toxic drugs, they become sicker, and then they all die."

Most of my experiences with clinical medicine were discouraging. Also, many of the patients who came to the hospital were suffering from physical diseases brought on by addictions — to nicotine, alcohol, drugs, even food — all of which I felt inadequately prepared to treat. We were never taught anything about addiction in med school. I felt we needed to know much more before we could approach such difficult problems as cancer, heart disease and addiction therapeutically. For me, the turning point came during my first year at the NCI. I decided to make a major switch in my career path and devote myself to science rather than medicine. I felt that I could help more people if I made breakthroughs in science instead of applying toxic therapies trying to kill all the cancer cells but succeeding only in killing my patients.

So I kissed goodbye the idea of becoming a rich American doctor, opting to become a sleuthing scientist. I told Lynn this would be good for us because I could set my own hours and spend more time with her and the boys.

Little did I know.

CHAPTER 2

Discovering Nirvana

These three terms would be fundamental in the new phase of my career as a scientist:

> *in vitro*, Latin, meaning in glass. For example, in a test tube or flask.
>
> *in vivo*, Latin, meaning in a living animal or human being.
>
> *clone*, Greek, *klon*, literally a twig; in biology, a population of cells derived from a single cell, whereby all individual cells are genetically identical, like twins.

While researching cancer and leukemia in the library, I realized that nobody knew what was wrong with a cancer cell, that is, how it differed from a normal cell. The only characteristic of malignancy was continuous proliferation. For me, it was logical to begin the search for the answer by first studying normal cells. I figured that we did not have a prayer to understand the abnormal behavior of a cancer cell until we knew how the growth of healthy cells was regulated. It's like a mechanic trying to fix a car in the shop; only by knowing how a car *should* function is it possible to fix the broken-down vehicle. This simple logic was game-changing, at least for me.

In my second year in Baltimore, I joined a laboratory focused on studying immunology, and especially lymphocytes, one of the four types of white blood cells (aka *leukocytes*, from Greek, *leuk*, meaning white, and Latin, *cyte*, meaning cell) responsible for immunity. I felt I had found my calling. I loved the intellectual, scientific, probing and questioning life. By comparison, medical life was dogmatic, tradition-bound, judgmental and frustrating. A heavy weight had been lifted off my shoulders.

The team I joined was small. My lab chief, Michael Mardiney, was studying how lymphocytes recognize and respond to viruses. At that time, viruses were thought to be the most probable cause of leukemia and other cancers. Mike was a native New Yorker, born and raised in Brooklyn, and he still had that distinctive Brooklyn accent. His medical training was in allergy, and he had done a research postdoc in immunology at the Scripps Clinic and Research Foundation in La Jolla, Calif. When he joined the NCI in 1967, he set up a lab focused on fundamental immunology. Len Chess, my lab bench mate from Columbia University College of Physicians and Surgeons, was from Brooklyn just like Mike. And as an aside, so was Anthony Fauci, another young physician-scientist who would go on to become director of the National Institute of Allergy and Infectious Diseases and chief medical adviser to President Joe Biden during the Covid pandemic. In 1974 I visited Tony's lab at the NIH and I came away with the impression that New Yorkers, especially from Brooklyn, must have an inside track on becoming immunologists. Len told me early on that as far as he was concerned the only things that really counted in life were the Yankees and the Knicks. He always wore a little mischievous smile. I found him amusing because he rarely said anything that did not have a hidden or a double meaning. It was as if he knew something that you didn't. I guess that's learned on the streets of Brooklyn. Len had already been in the lab for a year, and it was he, not Mike, who actually recruited me and became my day-to-day mentor. It was usually just the two of us in the lab because Mike stayed primarily in his office, planning experiments, and communicating with other immunologists, keeping his fingers on the immunological pulse.

I dove into the immunology literature, learning about it at the most basic level. Immunology was not a specialty in medicine then, and it still isn't. The closest subspecialties are allergy and rheumatology, both of which are focused on maladies caused by hyperactive immunity. Allergy deals with inflammation caused by noninfectious substances like pollen, while rheumatology deals with inflammation primarily of the musculoskeletal system caused by immune reactions to self (that is, autoimmunity). Therefore, therapies of these disorders are focused on immunosuppression. From the early 20th century until the 1980s, immunology consisted of observing the effects of the introduction of various substances into humans or experimental animals. Immunologists were at the macro level, trying to understand the immune system by observing changes in the whole body produced by foreign substances that provoked an immune response. It was like observing the ground from an airplane at 30,000 feet and trying to discern what people were doing. The way forward was obvious to me. It was necessary to reduce the size and complexity of the problem. I found that they had a term for this. It's called reductionist science.

The lab at the Baltimore Cancer Research Center was housed in a building separate from the hospital. Ours happened to be in the basement, without windows to the outside world. Nonetheless, going from the hospital filled with sick and dying cancer patients to the lab was like leaving hell and entering nirvana. The lab was outfitted for tissue culture with every conceivable device and machine necessary. It was one large room filled with a dozen typical lab benches with black soapstone tops. An array of equipment occupied space crammed on the benches and in between them. Critical for our work were laminar flow hoods ("clean benches"). They kept the lymphocytes from being contaminated with microbes by passing the air in the hood through a high-efficiency particulate air filter (HEPA) and blowing it over the sterile bench toward the investigator. We had tissue culture incubators adjusted to maintain body temperature (37°C, 98.6°F) and centrifuges of various speeds to precipitate cells at low speeds so they could be washed free from whatever media in which they were suspended, or to spin fluids at high speeds to clarify them from unwanted substances. We also had access to a special device that could freeze cells very slowly, by lowering the temperature 1°C each minute. It would thus take about two hours to go from room temperature to minus 80°C. Mike had adapted this device and procedure from the veterinarians who used it to freeze sperm for artificial insemination. After frozen, the lymphocytes were stored indefinitely in liquid nitrogen at extremely low temperatures (minus 320°F), which enabled us to retrieve and resurrect living cells from experiments gone by. The device would prove to be invaluable because we could obtain cells from the liquid nitrogen if our cultures became contaminated with microbes. We also seemed to have unlimited supplies for whatever was required to grow cells, including bottles of tissue culture media containing nutrients, salts and minerals in a walk-in cold room, and drawers and cupboards filled with plastic sterile flasks, pipettes and tubes. In short, we had everything we needed.

To obtain lymphocytes, we simply drew blood from our lab mates, a ready source fresh normal cells to study. Since we were physicians, drawing blood from each other was second nature. Had we been Ph.D. immunologists, this would have been a great deal more difficult since they never learned the skill of venipuncture. Consequently, they usually did not study human lymphocytes. Instead, they used mouse or rat spleen or lymph node cells. Even though rodents and humans diverged 80 million years ago, the Ph.D. immunologists essentially figured humans were just big rodents without tails. From their vantage point, it was not necessary to study humans. Our use of human lymphocytes represented another departure from the norm of immunology — and a distinct advantage. Because most card-carrying immunologists were not educated about all the diseases that humans encounter, their research seldom targeted questions related to diseases. And because the immune system had evolved with microbial infections supplying the force of natural selection, it seemed crazy to me to separate immunology from human disease and medicine.

Lymphocytes became excellent cells for studies on normal cell growth. Peter Nowell's 1960 experiments had established that lymphocytes normally circulated in the blood in a quiescent state, not proliferating. It was as a new assistant professor of pathology at the University of Pennsylvania that Nowell first showed their growth could be initiated by adding mitogens, agents that initiate mitosis (cell division) (2). We then monitored the changes in the cells over the next several hours or days. In perusing the literature, I found that most basic scientists studying the growth of normal cells usually used mouse embryo fibroblasts, which form fiber or connective tissue and could be grown continuously in culture media. From my reading, I learned these cells developed multiple genetic mutations when kept in culture for any length of time (months or years), so they could hardly be considered normal. To me, it did not seem like a good idea to base one's career on studying these cells because they were clearly genetically abnormal. Once again, my physician background influenced my career decisions. The road less traveled eventually made all the difference.

It was just the two of us in the lab. Len taught me how to isolate the leukocytes (white blood cells) from the erythrocytes (red blood cells), how to count them accurately using an electronic cell counter and how to suspend them in nutrient media at the proper cell concentration. Then we would distribute the cell suspension into various tubes that we would place in the incubators for several days. This was the beginning of the experiment, which would take us a morning to set up. The experiments consisted of control tubes to compare with tubes where antigens (defined as any substance that could stimulate an immune response) were added to stimulate proliferation. When Peter Nowell did his first experiments, he monitored the number of cells undergoing mitoses using a microscope, since he was a cytogeneticist (cell geneticist). One of my college biology professors commented that biology would not have progressed beyond the leaf-collecting stage without the use of radioactive isotopes, which allowed us to progress from actually seeing the cells by looking through a microscope to detecting molecules indirectly using sensitive new devices to detect the radioactivity. From chemistry, an isotope is an atom that has the same number of protons in its nucleus, but different numbers of neutrons. A radioactive isotope is critical because it is unstable and decays, giving off radioactive rays or particles, which can be sensed and counted in a device like a Geiger counter. By the time I entered the lab a decade after Nowell's report, it had become standard to measure the proliferation of lymphocytes by the amount of DNA synthesis. It was more accurate and faster to measure DNA synthesis than to count the number of dividing cells using a microscope, as Nowell had done. To measure DNA synthesis, we used nucleotides labeled with tritium (^{3}H), a radioactive isotope of hydrogen. Nucleotides are the individual building blocks (monomeric units) of the nucleic acid polymers DNA and RNA.

I started out with a project to stimulate the lymphocytes to proliferate using common viruses, such as mumps, measles and rubella (German measles). If an individual was immune to the viruses, either as a result of a previous infection or vaccination, there should be a detectable increase in the rate of DNA synthesis, which usually peaked after seven days of culture. The experiments provided a cellular basis to the immunity conferred by the childhood vaccines for mumps, measles and rubella and led to my first paper from the lab that was published in the new journal, *Cellular Immunology* (3). If an individual was immune to these viruses, the few naïve (never before exposed to a microbe) lymphocytes capable of responding to them originally had been expanded by their proliferation in response to the infection or vaccine. This seemed to me straightforward, but right away I thought it strange that it took a week for the cells to start proliferating *in vitro* after the addition of the viruses. It seemed like a long time. Little did I appreciate then that this observation was to occupy me for the rest of my career. We now know that the expansion of virus-specific lymphocytes is central to vaccination.

In 1957, an Australian physician, Macfarlane Burnet, had proposed the clonal selection theory to account for immunity. He went on to win the 1960 Nobel Prize together with Peter Medawar (4). Burnet came to his theory by studying immune responses to influenza virus infection after the 1918 pandemic. Scientists joined the fray to study flu that took 50 million lives worldwide, including more than a half million Americans. A century later, as of 2023, Covid has killed more than a million Americans, more than 6.5 million people around the globe, and it's not over yet. Burnet hypothesized that upon the first exposure to a microbe — or anything antigenic, for that matter — the few lymphocytes that could recognize it (he did not speculate how they did this) resulted in a selection of only these lymphocytes to proliferate. Later on, if a second exposure occurred, the expanded population of lymphocytes capable of reacting with influenza virus (that is, antigen-specific) was ready to respond rapidly and rid the body of the microbes before any symptoms appeared. This is key. The proliferation of lymphocytes is fundamental to the development of immunity and the reason that vaccination works. My experiments with mumps reflected the ease of detection of the expanded number of mumps-reactive cells resulting from the MMR (mumps, measles, rubella) vaccine for infants, compared with the unvaccinated, who had only a few antigen-reactive cells below the level of our detection methods.

As a result of these experiments, I realized that it would be possible to study the mechanisms responsible for activating normal cell growth using lymphocytes as a model. However, as a consequence, I had fallen down a rabbit hole into a new realm of cells and molecules that comprise the immune system. By default, I would become an immunologist, not a cancer researcher.

One Sunday afternoon I went into the lab to check on my experiments. I turned on the radio to hear the Cleveland Browns football game and was working away when I heard opera music wafting from the lab next door. It turns out that another New Yorker, Bob Benjamin from NYU, was also working that day. I walked over and started up a conversation. I figured this guy must be a little different than what I, coming from Ohio, was used to. He certainly was. Bob was the only child of a couple of academics born when they were in their 40s. They raised him in New York City, essentially as if he was an adult. He was well-schooled in the finer things in life, such as opera, wine and food. We became fast friends as did our wives. For Lynn and me, this was the beginning of us becoming Francophiles, which was to be life-changing as will become evident. You might expect that Bob was a little nerdy, growing up as he did. Not at all. He was a charming, enthusiastic teacher of all he had learned. And we were enthusiastic students.

Bob invited us to join a wine club that met in the wine cellar of one of the finest restaurants in town, the Chesapeake. The restaurant's young sommelier introduced us to wines from all over the world. I quickly became obsessed with learning about wine. I bought Alexis Lichine's *Encyclopedia of Wine and Spirits* and read it cover to cover. Haut-Brion, one of only five "first growths" of Bordeaux vineyards, was the last tasting of the year and became fixed in my memory. Right then I vowed that I would visit as many of the French vineyards as possible.

This world of wine led to another delightful turn of events. One night Nancy and Bob Benjamin had us to their apartment for dinner, and it proved to be of special importance for Lynn. She was so impressed by the meal that she asked Nancy for the recipe. Nancy said that it was from Julia Child's *Mastering the Art of French Cooking, Volume I.* On the way home, Lynn announced, "If Nancy could make such a delicious dinner by following Julia, then I can too." Whereupon she borrowed the cookbook from the library and set herself to cooking everything in it, beginning on page one, just as it would be portrayed by the movie *Julie & Julia* four decades later. "Julia was responsible for maintaining my sanity when our two toddlers were my only sources of conversation," Lynn recalled. Kelly was two months old and Kerry was just two years old when we moved to Baltimore. Our family benefitted immensely from her new focus. The boys grew up on Julia's cuisine and developed sophisticated palates just like I imagine Bob Benjamin had. Of course, I also benefitted tremendously. We were all learning to live *comme les Francais*.

In the lab, Len was interested in understanding *adjuvants* — a term used for any substance that boosts the immune response. As I became proficient and comfortable with techniques in the lab, I helped Len with his project. In the early days of bacteriology in the late 19th century, Dr. William Coley reported that injections of bacteria directly into tumors could cause their complete disappearance (5). A half century later, in the 1940s, investigators explored ways to improve immunization and discovered that bacteria could be used as adjuvants to enhance vaccine efficacy when injected along with a vaccine (6). But no one knew how adjuvants worked, how they boosted immune responses. As a model, Len explored their effects using synthetic RNA (ribonucleic acid) molecules designed chemically to mimic viruses. These molecules were simpler than using a whole microbe. Len had already found that if he added a molecule called polyAdenine-polyUridine (polyAU) double-stranded RNA (dsRNA) to his lymphocyte cultures, he could achieve a marked increase in lymphocyte proliferation. Somehow, the cells recognized that the dsRNA molecules were foreign. We naturally wondered how these dsRNA polymers functioned as adjuvants. A clue came from a recent report that other white blood cells called macrophages were necessary for lymphocytes to proliferate. The term *macrophage* is also derived from Greek that literally means "big (macro) eater (phage)." Joost Oppenheim at the NIH had shown a few years earlier that lymphocyte proliferation in response to antigenic stimulation is dependent on the presence of macrophages, but no one knew precisely how they contributed to lymphocyte growth (7).

We proposed to see whether the polyAU stimulated either the macrophages or the lymphocytes. It seemed like a simple enough question. Mike Mardiney, our lab chief, told us that we were spinning our wheels, a phrase I have recalled often since then. Bucking the boss, I followed the methods in the literature to separate the macrophages from the lymphocytes, which entailed placing the leukocytes for 30 minutes in a plastic flask. The macrophages would settle and spread out like fried eggs and attach to the plastic while the smaller round lymphocytes remained in solution. We could then rinse off any remaining lymphocytes with more tissue culture medium (a salt solution plus essential nutrients). We treated the two separated cell populations with the polyAU, culturing them separately and together, and monitored them for lymphocyte proliferation. The results were clear-cut: the polyAU stimulated the macs, as we liked to say, but not the lymphocytes. This resulted in my second publication (8). The mystery was how the polyAU promoted the macrophages to "help" the lymphocytes. The answer was not immediately evident, but the advantage of the *in vitro* culture system was obvious.

My first year in the lab made me realize it was going to take just as long to become a scientist as it had already taken me to become a physician. This awakening culminated in a fit of depression one evening after a trying day in the lab. I was so down on myself that I told Lynn, "I don't think that I'm creative enough to become a scientist. Try as I might, I can't think of any new questions to investigate. Meanwhile, Len seems to be always brimming with new ideas." Lynn rolled her eyes. "Give it a little more time," she advised me. "You're new to this. You've only been at it for a few months." Excellent advice. The next day after a good night's sleep I regained my self-confidence and recommitted myself to a career in science.

Falling back on practicing medicine would have been admitting failure. I was determined to not let that happen. As time went on and I gained experience, I realized my background as a doctor prepared me well for science. It gave me a broad knowledge of normal human physiology, the scientific study of a living system. I had also learned about pathophysiology, the basis for disease. I could build upon that branch of science by extensive reading, thinking and discussions with other researchers. I never again distrusted my capabilities, and I came to realize self-confidence is one of the most important characteristics of success in science. First, believe in yourself.

I knew I needed to find a mentor who could shepherd me through my postdoctoral training and the business of science. From the scientific meetings I attended, it was clear there were different power groups and young people served as apprentices with the most famous scientists to become members of an elite corps. I could have stayed at the NCI in Baltimore with Mike, but Len was leaving at the end of his two years there. I felt it best to move on. Mike taught me a great deal, but he was not in the lab much. I needed hands-on supervision. I applied to postdoctoral research programs all over the country, but primarily in the East and was accepted into a research training program at Dartmouth Medical School under the guidance of Ross McIntyre, a physician and chief of the Division of Hematology. McIntyre's claim to fame was that he was second, after Peter Nowell, to publish in the early 1960s that lymphocytes could proliferate in culture (9). The problem was that he had published little recently in the field. Mike counseled me not to take the job, but I didn't listen. One might say that I was headstrong. That's a shocker. Lynn was thrilled to be leaving Baltimore and our small apartment with our two very active little boys to take up residence in Hanover, New Hampshire, a quintessential college town with some 5,000 residents and 5,000 students. With the town Green surrounded by 18th and 19th century white brick buildings of the campus, it is a New England setting right out of a Currier and Ives print. We borrowed the down payment from Lynn's father to purchase our first house, and we were ecstatic to begin a new adventure.

When I reported for work at Dartmouth on July 1, 1972, my mentor informed me there was good news and bad news. The bad news was that his recent NIH research grant renewal application was not re-funded, so he had no money for me to conduct the experiments we planned. The good news was that he still had enough money for my salary. I had left a situation where money was no object to join one where basically there was no money. You could say that I was not very street smart. Yes, I heard Mike in my head telling me, "I told you so." Needless to say, Lynn and I were more than a little upset over this state of affairs. Right then and there, I vowed to never treat my younger trainees like that. I felt that Ross should have given me at least a little advance warning. I pledged to adopt the golden rule for my subsequent behavior in the lab as a mentor, assuming that eventually I would attain such a role.

Normally, postdoctoral research fellowships are set up two years in advance. Dreading two unproductive years in the lab, I sat down and wrote letters to search for a postdoctoral fellowship that I could begin the very next year. I just couldn't see cooling my heels for two years. Once I had made my decision to become a scientist, I was eager to begin. I wrote especially to European labs because I had always wanted to study in Europe. I was accepted for the next year by Dr. Georges Mathé at the Institut de Cancerologie et d'Immunogenetique in Villejuif, one of the *banlieues* (suburbs) on the southern outskirts of Paris. This was great. Under the prior tutelage of Bob Benjamin, I had come to consider France the epicenter of culture with its emphasis on food and wine and history of literature, philosophy, art and music.

Our house in a new suburban development in the neighboring town of Lebanon was a variation of the New England saltbox with the ground floor serving as a playroom and garage, while all living was done on the second floor. A deck off the kitchen at the back of the house abutted a small hill. For the first time, we had a large yard. Lynn could just open the door and let our two little boys run around to their hearts' content. We met neighbors with young children right away, which was to be key for our family's happiness. When I was not occupied in the lab, I took advantage of what Dartmouth had to offer. I enrolled in French 101, hoping it would help me in Paris. Almost all my fellow students were freshmen, 17 or 18 years old. They looked upon me as an old man at 30. I was not one of the best students in the class. Luckily, the professor was kind to me as I struggled with both grammar and pronunciation. It was humbling, given my senior status and prowess as a medical student. I studied late into the night to keep up with the class, which moved rapidly, as most of the students had already had several years of French.

Ross's Dartmouth lab was in the Mary Hitchcock Memorial Hospital, separate from the rest of the med school research labs. Ross said that he did not want to be too far away from patients, which should have told me a lot about where his interests were focused. The lab rooms were large enough and adequately equipped for tissue culture, but they were antiquated compared with the NCI labs. Nonetheless, I settled in and started acquainting myself with the two techs and looked to see whether I could duplicate the techniques I had learned in Baltimore at Dartmouth.

Having secured a position for the following year, I was determined to learn as much as I could from the lab at Dartmouth. Ross had a clinical research project focused on polyinosinic:polycytidylic acid (polyIC), another synthetic dsRNA molecule similar to polyAU, on which Len Chess and I had worked together. Like polyAU, polyIC had been found to be a potent adjuvant, as well as an inducer of *interferon*, which was defined as an "antiviral activity". Interferon had been first identified in 1957 (10) in the culture media of cells infected with viruses. Interferon mysteriously protected, or interfered, with the viral infection of neighboring cells, thereby restricting the viral spread. It came to be regarded as a first line of defense the immune system used against invasion by viruses. It was also found to be a target of successful viruses that come up with ways to block infected cells from producing interferon. For example, smallpox and measles viruses both block the early production of interferon, thereby permitting the virus to spread without causing symptoms. The coronavirus has adopted this same trick so that after infection the virus can replicate unimpeded for a week to 10 days without causing the symptoms of inflammation. These appear only when the adaptive immune response kicks in. The fever and malaise of Covid occur as the immune system attempts to rid the body of the virus, the first alert to the person infected that something is amiss.

There was a growing excitement about interferon as a potential anticancer agent in the 1970s because it inhibited the proliferation of cancer cells *in vitro*. The problem was that no one had enough interferon to try it as an anticancer agent in humans. Interferon was still only a measurable "activity" that had not yet been identified to be due to an actual molecule, comprised of atoms bonded together that could be purified and mass-produced, put into bottles and administered to patients. Before interferon could become a therapeutic agent, first it would be necessary to identify a molecule responsible for the antiviral activity, second to separate it from other molecules (purify it) and third manufacture it in large enough quantities to administer it to multiple patients *in vivo*. All three steps depended upon improvements in the technologies of biochemical and molecular research that had yet to happen. As another option, if it was possible to harness interferon *inducers* such as polyIC as a medicine, it might open up a new whole new form of therapy for viral infections and cancer — immunotherapy, using adjuvants as interferon inducers — and provide an avenue to bypass all the painstaking research needed to use interferon itself.

To circumvent the lack of interferon to administer to cancer patients, Ross was conducting polyIC dose-finding, phase one clinical studies in humans to determine its toxicities. He was keen to see if polyIC could induce the production of interferon in humans. I was assigned to give the polyIC intravenously to the volunteers and monitor their vital signs. We found that the injection of polyIC induced fever with a characteristic time-course, peaking after about 18 hours. Ross had set up assays (measurement methods) in his lab to determine interferon concentrations in the plasma, the liquid portion of blood, of subjects after the administration of polyIC. Learning how the interferon assay was performed was to prove important for me in the future. It consisted of culturing human fibroblasts, cells that form fibrous connective tissue like ligaments and tendons, together with serial dilutions of interferon samples for a few days, then challenging the fibroblasts with viruses that would usually kill the cells. Because the interferon blocked virus infection and replication, we could plot a curve of the amount of cell killing on the Y-axis and the dilution of the interferon sample on the X-axis. The samples that could be diluted the farthest before losing the viral interference activity contained the highest concentration of interferon, making it a *quantitative* assay. This was important because then we could compare the interferon concentrations in different samples. All of this may seem like I had taken off on a tangent. As you shall see, it turned out that interferon is in a family of similar molecules now known as *cytokines* (from Greek *kine*, meaning move, as in kinetic). Measuring interferon activity would become important indeed as I was destined to spend my career as one of the pioneering scientists riveted by cytokines.

While a technician performed the interferon assay, I did lymphocyte stimulation assays using the techniques I had learned in Baltimore, hoping we could use them as an indicator of immunity. We found that polyIC induced the appearance of plasma interferon activity that exactly coincided with the fever curve. Interferon and additional cytokines yet to be discovered were subsequently shown to cause fever, one of the cardinal signs of inflammation. Our report in the *Journal of the National Cancer Institute* was among the first to reveal that adjuvants like polyIC could produce cytokines like interferon simultaneously with fever in humans (11). We also found that interferon inhibited rather than enhanced the proliferation of normal lymphocytes (12). These results were counterintuitive to my concept from the results of our experiments in Baltimore showing that adjuvants like polyAU and polyIC augmented, rather than inhibited, lymphocyte proliferation. This was my first glimpse into the complexity of the immune response that I was to confront again and again as I tried to understand it and harness it for therapeutic purposes.

Time flew by. Our year in Hanover was over. Lynn had researched the costs of transporting our family with all the paraphernalia necessary for toddlers across the Atlantic. The least-expensive way she found was on an ocean liner, which let us take unlimited luggage. On June 25, 1973, the day John Dean testified before the congressional Watergate hearings about "a cancer growing on the presidency," my family embarked on the SS France in New York bound for Le Havre. We were happy to be leaving the U.S., given the sad state of the government. I was excited to begin a new research adventure and to meet Professeur Mathé.

CHAPTER 3

The French Connection

Georges Mathé was 18 years old when the Nazis invaded Paris in 1940. Instead of continuing his education, he spent the war fighting with the Resistance and was 29 when he graduated in medicine from the University of Paris in 1951. After hospital training, Mathé elected to join *"le pere de l'hematologie francaise,"* Paul Chevallier, to focus on leukemia at Hôpital Saint-Louis. Then, building on this initial clinical experience in hematology caring for patients in the hospital in the center of Paris, Mathé came to the United States for a postdoctoral fellowship with pioneering oncologists David Karnovsky and Joseph Burchenal at Memorial Sloan Kettering hospital and cancer research center in New York City. During these years, at the dawning of chemotherapy, Mathé became skeptical that injecting toxic chemicals could be a cure for leukemia. He concluded that there must be a better way (just as I would conclude a decade later). It was perhaps fortunate for Mathé that Burchenal was experimenting with other approaches.

One of them consisted of transplanting bone marrow, the source of blood-forming cells, especially immune cells, into mice with leukemia. Mathé encountered a whole new idea on cancer treatment, and his contemporaries training at Sloan Kettering were to influence his future experimental directions. Back in Paris, under the tutelage of Chevallier and Jean Bernard, Mathé reported results from experiments in 1958 and 1959 examining the effects of bone marrow transplants as treatment for mouse leukemias induced by RNA tumor viruses. These are now termed *retroviruses* because the information in their RNA genome, which contains the genetic code, or information, could be "reverse-transcribed," or changed from RNA into DNA, with the genes of the retroviruses inserted into cellular DNA. Retroviruses, known for many years to cause leukemia in animals, are now familiar to many people because of the retrovirus HIV, human immunodeficiency virus, that causes AIDS.

Inbred mice (incestuous breeding over 20 or more generations) were developed in the 1930s at the Jackson Laboratory in Bar Harbor, Maine, making possible many genetically different mice available for study. Mathé found that a bone marrow transplant could successfully eradicate leukemia in mice, with the caveat that he injected only a small number of leukemia cells.

Mathé was investigating bone marrow transplantation to treat mouse leukemia when an unfortunate, but rare opportunity presented itself. Six Yugoslav physicians suffered lethal radiation from an accident at a nuclear power plant. Faced with their sure death, Mathé and his colleagues in Paris attempted to rescue the radiation-damaged blood-forming systems of these physicians using bone marrow transfusions from unrelated individuals. In 1959 Mathé reported the world's first human bone marrow transplant (13). The gravely ill patients died of what came to be known as graft vs. host disease. The grafted immune systems of the donors' bone marrow rejected the recipients' normal cells and tissues. Mathé was credited with the first demonstration of graft vs. host disease, showing that it was an immune reaction of the transplanted donor against the recipient.

In 1961 at the age of 39, Mathé was offered the opportunity to head the Hematology Department of the Institut Gustave Roussy, an internationally recognized cancer research and treatment center in Villejuif. Buoyed by his initial bone marrow transplantation in humans and the results of his mouse leukemia experiments, Mathé pressed on to try to treat human leukemia by transplanting bone marrow. In 1963 he reported the world's first successful bone marrow transplantation between unrelated individuals, treating a 26-year-old physician suffering from acute leukemia with a mixture of blood and bone marrow cells from six family members (14). This was before selection of compatible donors was discovered via tissue typing. After eight months, the patient's blood was repopulated with red blood cells of one of the donors. In addition, antibodies (molecules that bind to antigens) of the same donor type appeared in the recipient's blood and a skin graft from the donor was accepted, rather than rejected. All of this indicated that the recipient's blood-forming and immune systems had been replaced by the donor's cells. Nowadays, we "type" the donor and recipient so that they match each other as closely as possible to try to reduce the incidence and severity of the graft vs. host syndrome.

Mathé wrote in the *British Medical Journal*, "The fact that our patient, previously in a phase of relapse with resistance to most forms of treatment, had at the time of writing eight months of complete remission, constitutes in itself a remarkable result" (14). After 20 months, there was no evidence of leukemia (14). With his report of the first successful bone marrow transplant in man, Mathé conjectured that bone marrow transplantation had had an anti-leukemic effect, graft vs. leukemia, similar to the graft vs. host effect. In other words, it appeared that the lymphocytes of the graft rejected the leukemia cells of the host. With these reports, Mathé's group not only achieved the first successful bone marrow transplant, but was the first to describe both graft vs. host and graft vs. leukemia as well.

To put these remarkable results into historical perspective, it was not until almost 30 years later in 1990 that the use of bone marrow transplantation to treat leukemia and other blood disorders was recognized by the Nobel Prize, and it is noteworthy that Professeur Mathé was not a recipient. The prize went instead to E. Donnall Thomas, a physician from Seattle educated at Harvard who capitalized on Mathé's reports, confirming, extending and popularizing them. Evidently, the Nobel committee felt Thomas did more to bring bone marrow transplantation into mainstream clinical medicine than had Mathé despite being first.

Georges Mathé was a rising star and in 1964, just a few years after his postdoc in New York, he was made director of the Institut du Cancerologie et d'Immunogenetique, a new research center in Villejuif. This was an obvious challenge to America's National Cancer Institute, which had been established in 1937 by President Franklin Roosevelt. The new French cancer center was intended to thrust French cancer science onto the international scene. In 1967 the Faculte de Medicine in Paris awarded Mathé the Chaire Cancerologie Experimentale, anointing him with the appropriate credentials of his status within the French academic establishment (15).

Only two years later, in 1969, Mathé's team rocked the cancer world with a report of a new immunotherapy for acute lymphoblastic leukemia (16). Added to standard chemotherapy, which used chemicals that killed cells, essentially poisons, this immunotherapy intended instead to boost the immune system to kill the leukemia cells. In the report, published in 1969 in the pre-eminent British medical journal, *The Lancet*, Mathé explained that there could be two kinds of immunotherapy, nonspecific and specific. Nonspecific immunotherapy meant the therapy was not purposefully directed towards promoting the immune system to kill the tumor cells, but rather to nonspecifically boost the overall function of the immune system. Nonspecific therapy used molecules, much like polyAU and polyIC, to boost immune responses nonspecifically, as adjuvants. Mathé had also tested various bacteria for adjuvant activity. He found that bacillus Calmette-Guerin, a tuberculosis vaccine developed from bacteria at the Institut Pasteur in the 1920s, was particularly effective (17). It might seem strange that Mathé chose to use nonspecific immunoadjuvants, especially a live microbe to boost the immune system. In fact, his decision rested on years of precedents. Immunotherapy was first attempted by William Coley, who used live bacteria, then heat-killed bacterial extracts, hoping to boost antitumor immune responses in his patients suffering from malignant bone tumors (5).

By the time of Mathé's experiments, it was well-known that TB organisms mixed with oil (thought to retard removal of the vaccine) boosted immune responses. Subsequently, animals were routinely vaccinated using an antigen mixed with dead TB microbes and oil.

1973: Professor Mathé at the height of his celebrity.

This had come to be termed "complete Freund's adjuvant," named for New York immunologist Jules Freund who worked with guinea pigs using horse sera as an antigen in the 1940s (18). While it is still considered ethical to use animals (especially rodents and in particular mice, which are a favorite experimental animal) before or instead of humans for the life sciences, experiments using primates are now facing increased scrutiny as more and more evidence accumulates that animals are not very different from humans. As an aside, E.O. Wilson in his landmark book, *Sociobiology*, shows how all animals, even insects, have evolved complex societies with behaviors remarkably similar to our own (19). It may well evolve that all animal experiments will one day be considered unethical.

While the idea to treat cancer indirectly by boosting the immune system rested on historical experience, it was still too revolutionary for cancer specialists especially chemotherapists, to accept. As far as traditional medicine was concerned in the mid-1960s, the immune system was an enigma. Most oncologists felt it was preposterous to try to boost the immune system nonspecifically to kill cancer cells because the cancer cells were derived from *self*, whereas the microbes were clearly *nonself*, and everyone thought they knew that the immune system only attacked what it recognized as nonself, like bacteria and viruses, as well as allergens like pollens.

As specific immunotherapy, Mathé injected irradiated leukemia cells taken from multiple other leukemic patients. This hypothetical tumor antigen specificity was based on Mathé's mouse leukemia experiments, which showed that *tumor-specific* antigens could be detected by immunizing mice with tumor cells, then testing for resistance to transfers of the same tumors. He hoped that human leukemia cells also expressed such tumor-specific antigens. If so, leukemia cells could also be used as therapeutic vaccines, rather than just prophylactic, or preventative, vaccines. Pasteur had introduced the concept of not only prophylactic vaccines, but also therapeutic vaccines that could be used to treat rabies after the bites from rabid dogs had occurred (20). The other crucial aspect of Mathé's novel approach, based on his mouse experiments, revealed that immunotherapy had to be delivered when the number of tumor cells inoculated was low, under 100,000 cells. In humans, this meant it made sense to reduce the number of leukemia cells first with chemotherapy, before the immunotherapy was applied.

Mathé's team began a clinical trial in 1965 to test a variety of immunotherapies in subjects with acute lymphoblastic leukemia, the most common form of childhood leukemia. This trial proved to be of prime importance for the development of cancer immunotherapy — and for my own research. For the nonspecific immunotherapy, bacillus Calmette-Guerin was given every fourth day during the first month and once a week thereafter. To administer the vaccine, 20 scalpel cuts were made on the skin, usually on the patient's back, each two inches long arranged in a square. On each square, 2 milliliters (about one half teaspoon) of a suspension containing 75 milligrams per milliliter of living bacteria was applied. This produced an open wound, which would scab over and become red, hot, swollen and sore (the cardinal signs of inflammation) before sloughing off after a week or two. The specific immunotherapy consisted of weekly injections into the skin of 40 million leukemia cells taken from other leukemia patients.

Thirty subjects were entered into the trial after chemotherapy-induced complete remission, defined as the disappearance of leukemia cells in the blood and bone marrow. Ten subjects were randomized to receive no immunotherapy and serve as controls, while 20 subjects received immunotherapy. By January 1, 1969, all 10 controls had relapsed within five months after stopping chemotherapy. Of the 20 immunotherapy subjects, 12 had relapsed, but eight (40 percent) remained in remission after the chemotherapy was discontinued. These results generated great excitement in the oncology community. Notably, it appeared that nonspecific and specific immunotherapy yielded equal efficacy. This allowed subsequent investigators to use either one rather than both. Mathé continued to treat his long-term immunotherapy responders for years. In 2001 he reported that seven of these initial eight subjects had not relapsed by their twelfth year. They were cured, for all intents and purposes, a remarkable result reminiscent of Coley's cures of a century earlier (21)

It was against this background that I was excited to be joining Mathé's institute, just four years after his remarkable immunotherapy publication. Mathé warned me not to come until September because most people in France took vacation for the entire month of August and the institute would be *inhabite*. However, my postdoctoral fellowship at Dartmouth was to end in June, the usual termination of academic medicine appointments in the States. This practice stemmed from the stipulated residency periods in hospitals. After a wonderfully peaceful transatlantic voyage, we arrived in Le Havre and took the boat train to Paris, which gave me my first glimpse of the French countryside. I marveled that the Old World landscape was no different from our New World at home. I expected it to be different in some way, you know, older. Upon arrival in Paris, one of the institute's postdocs picked us up at the Gare du Nord. En route to Villejuif, he gave us a whirlwind ride through the center of the City of Lights, which we found to be true to its name, with colorful bright lights everywhere, and shocking, noisy, fast-moving traffic.

With the labs *en vacances*, we spent the rest of the summer exploring Paris and setting up our new home. The apartment provided by the institute had two small bedrooms and a living-dining area. The tiny kitchen had a refrigerator a third the size of an American one. This space reduction took more than a little adjustment for our family, especially Lynn. The idea of the four of us spending the next few years crammed into this apartment, even smaller than our Baltimore abode, was depressing. I had already decided that I might stay longer than the usual two postdoc years if everything went well, so our living situation made this dream seem like a potential nightmare.

This portrait was taken from the passport photo used when we traveled to France on the S.S. France.

We lived on the fifth floor of a six-story building with shops on the ground floor. There was a *boucherie de cheval* (horsemeat butcher), whose eponymous product we found to be not too different from beef. We loved the wonderful *boulangerie-patisserie* next door for its bread, quiches and, of course, croissants and baguettes right out of the oven. Each morning we'd give Kerry and Kelly a few francs and they would take the elevator down and shop all by themselves for our *petit dejeuner.* The proprietor of the *boucherie de cheval,* M. Dumeseaux, became a good friend to Lynn and the boys, teaching them the French words for what he sold. Here again Julia Child's cookbook proved its value because Lynn would write down the French food names to give to the shopkeeper. In the beginning, we ate a lot of *tranches de jambon* (sliced ham). The best ham for making a *jambon-beurre* sandwich is called *jambon de Paris,* which is made by French butchers. It is wet-brined and slow-cooked to retain a lot of that moisture and soak in the flavors of the broth it's cooked in. Julia's book gave the food terms and measurements in both metric and English equivalents, which was also certainly helpful as Lynn continued to experiment with *la cuisine francaise*.

One of the characteristics of apartment living that I didn't understand when we arrived was that people shuttered tight their windows at night. I quickly got the message because the noise from the street was absolute cacophony. It's only now that Paris has instituted new laws to fine noisy motorists. I could not figure out how anyone could sleep through the racket. Our building was on a corner of Route Nationale 7 one of the main arteries, designated by Napoleon, leaving Paris south to the Cote d'Azur. All night long trucks would screech to a stop at the traffic light, rev their engines while the light was red, then roar upon takeoff to achieve maximum speed when the light turned green. A neighborhood *brasserie* was on the other corner, where a side street abutted the Nationale 7. I was amused to watch truck drivers pull up at a red light with one wheel on the sidewalk, run inside and throw down a shot of cognac, then run back and be on their way when the light turned green.

We quickly discovered the many parks and playgrounds of Paris. One of the quite civilized aspects of French playgrounds was the ever-present nearby café kiosk where one could enjoy delicious patisseries and a café or even a glass of wine, definitely not a feature found around an American playground. We bought toy sailboats for the boys, and they sailed them in the fountain in the center of the Jardin de Luxembourg while we sat at a kiosk with our wine. During these sojourns, I watched the behavior of the French, fascinated by their characteristic body language, facial expressions and gestures. I practiced and tried to replicate them. Lynn said that with this, as with many things I did, I tended to exaggerate "slightly."

In September after our two-month *vacance*, I sat down with Professeur Mathé for the first time to discuss possible research projects. His ongoing clinical project focused on developing laboratory assays to monitor the immunological status of the patients undergoing immunotherapy. This project had been initiated by two previous American postdocs. They had recommended me as the perfect person to continue their research. Mathé said I could participate as a clinical liaison, administering skin tests and ferrying blood from the clinic to the technicians in the lab. I said, "That might be OK for a start, but I also want to delve into more basic projects." From my experience at Dartmouth performing lymphocyte stimulation assays on patient samples after polyIC treatments, I knew these kinds of assays could not easily quantify changes in immune status longitudinally, that is, over time. There was just too much variability from one day to the next, so I was skeptical about the project. Based on my experiments studying polyIC at Dartmouth and polyAU at the NCI with Len Chess, I wanted to explore the mechanisms by which adjuvants could boost immune responses. This seemed to be where the action was, especially if immunotherapy was so effective in the treatment of leukemia. I knew the obvious next question to explore was the effect of adjuvants on macrophages that Len Chess and I had shown were necessary for the effects of polyAU to enhance lymphocyte proliferation. Moreover, I had some ideas as to how to proceed.

Just over three years earlier, it had been reported that macrophages could make "activities" or "factors" that would substitute for the requirement of actual macs themselves in lymphocyte proliferation assays (22). For example, lymphocyte activating factor (LAF), derived from macrophages stimulated by bacterial lipopolysaccharide (LPS), had just been described in two papers in the *Journal of Experimental Medicine* (23, 24) as a soluble factor that boosted lymphocyte proliferation. "Factor" was the word used to indicate a substance that had not yet been defined at the molecular level. The term itself was a way to hedge because it was not clear exactly what the substance was. I figured we needed to discover the molecular basis for this substance. My notion triggered a guffaw from Mathé. He informed me this was way too complicated a project for a new *stagiaire postdoctorale* such as myself. His reaction struck me as strange. I thought the molecular mechanism or mechanisms, whatever they were, could be central to how the immunotherapy of leukemia worked. To proceed further in that immunotherapy, I thought we needed to know exactly what adjuvants were and how they functioned. As a physician, schooled in the pharmacological action of drugs, it seemed obvious to me.

Mathé spoke impeccable English with a soupcon of a French accent, reminiscent of Maurice Chevalier. Delightfully charming and affable, he was quick-witted and found amusement in many of the people with whom he came in contact. At this stage in his career, he was traveling a good deal, but when he was home, there would be a *grand dejeuner* for all of us at the institute. With 100 researchers seated in a large square in the dining room, Mathé entertained the gathering during the luncheon with chatter about the people he met on his travels and the science he discussed with researchers around the world. I made it a point to sit next to him so I could ask him to translate something I did not understand, especially the French slang. One particular *histoire* had to do with William Terry, the chief of an immunology lab at the NCI. When this American scientist came to visit, Mathé drove to Charles de Gaulle Airport, north of Paris, to meet him and personally take him down to Villejuif. When they reached the center of the city, Mathé stopped at Notre-Dame to give his passenger a quick view inside the famous cathedral. Terry returned to the car in just a few minutes, shrugging his shoulders and telling Mathé, "If you've seen one church, you've seen them all." This story prompted a burst of laughter and applause throughout the hall. Of course, the enlightened French had reduced the power of the church with the French Revolution two centuries earlier, unlike the Americans still under the yoke of the Puritans.

The institute occupied a hill overlooking an academic complex including the Hôpital Paul-Brousse, the Institut Gustave Roussy, and the Centre National de la Recherche Scientifique, which was equivalent to the U.S. National Science Foundation. Looking north from the institute on a clear day, one could see in the distance the white spires of the 19th century Basilique du Sacre-Coeur de Montmartre on the summit of a butte, the highest point in Paris. We also had a great view of Sacre-Coeur from the kitchen window of our fifth-floor apartment. The institute comprised a half-dozen newly constructed single-story buildings, each devoted to a different aspect of hematological or immunological research. Over the course of the first few months, I spent my mornings in the clinic, ferrying the blood samples of the immunotherapy patients to the lab techs. I observed firsthand the scarification technique, whereby the two-inch-long cuts were made with a scalpel into the patient's skin, followed by the application of bacillus Calmette-Guerin onto the open wound, by no means a trivial procedure. It was traumatic for both the patient and the doctor. In the afternoons, I explored the different labs to get a feeling for what other possibilities might be there for me. Most were devoted to experiments with mice, investigating bone marrow transplants, trying to see if it was possible to separate graft vs. host from the graft vs. leukemia effects. Others were studying various immunotherapies to boost immune reactivity (notably polyIC and polyAU) to compare with bacillus Calmette-Guerin and discern which was most effective (25). But no one was studying how nonspecific immunotherapy actually worked, so I could not identify a more basic project to pursue. Unbeknownst to me, the two previous American postdocs had sullied the waters at the institute by behaving as if they were better than their hosts. To the French investigators, they exemplified "the ugly American." I was looked upon with skepticism, just another of the same ilk.

After a few months, I met a fellow American, Torgny Fredrickson. Torg was on sabbatical from the University of Connecticut and an internationally recognized mouse hematopathologist, that is, a specialist in the microscopy of blood disorders. He was particularly interested in mouse erythroleukemia. Erythroleukemia is a misnomer because leukemia is defined as cancer of the white blood cells, or leukocytes, but *erythro* (the prefix means red in Greek) designates a cancer of red blood cells, or erythrocytes. Such is the coining of terms by scientists, not linguists. Torg was collaborating with a group across town studying erythroleukemia at the Curie Institute at the University of Paris at Orsay. After the student riots of 1968, the University of Paris was broken up into satellite institutions with many students farmed out to the *banlieues*. The unit at Orsay was focused on the sciences, while the humanities were distributed to other branches. By this time, it was known that retroviruses could consistently cause mouse erythroleukemia. One had been identified at Memorial Sloan Kettering by Charlotte Friend in the 1950s (Friend leukemia virus) and another by Frank Rauscher in the 1960s at the NCI (Rauscher leukemia virus). The Curie Institute investigators were experts in both Friend and Rauscher virus-induced erythroleukemia.

Torg and I talked over coffee and later lunch, two expats sharing observations about the quirks of Parisians. We quickly became close friends. Torg was about 40 years old, of Norwegian descent, tall and lanky. We were like Mutt and Jeff, since he was well over six feet and I am only five four. He had a dry but zany sense of humor and was like a novelist, noticing and describing the eccentric mannerisms of the French.

In our free-ranging scientific discussions, Torg educated me about erythroleukemia and the Friend and Rauscher viruses, and about erythropoietin, an activity discovered to regulate the production of red blood cells in the bone marrow. This suffix is another from Greek (*poietin* means to make or produce) and thus, its activity stimulates *erythropoiesis* (red blood cell production). My time would not be wasted working on erythropoietin, because years later this activity was found to be in the cytokine molecular family, to which the interleukins also belong.

Combining our expertise, Torg and I devised an assay for erythropoietin based on the one for interferon I had learned at Dartmouth. Recall that the interferon assay was quantitative, meaning it allowed us to measure the amount of interferon in a sample. This would serve as a model for my later development of the first quantitative assay for T cell growth factor (26). *Assay* can be either a noun or a verb. It is from the Old French and connotes a *measurement* (noun) or *to measure* (verb). To construct a bioassay (noun), we needed a supply of living cells responsive to erythropoietin. Torg informed me that mouse fetal liver cells were a rich source of erythropoietin-sensitive cells. In the first trimester of pregnancy the developing liver contains most of the blood-forming cells. Torg was able to obtain essentially unlimited quantities of pregnant mice from the Curie Institute. To determine when a pregnancy occurs, an experimenter placed a male and two females in a cage overnight. I don't know why this was a male-female-female (MFF) mix. The next morning one would check for vaginal plugs of seminal fluid obliterating the vaginal opening. If a female mouse was "plugged," it was separated into a private cage, dated and observed for pregnancy. Torg would arrive in the lab with a big smile and a mouse cage if he was successful in obtaining a pregnant mouse.

We removed the embryos at the end of the first trimester (10 days). At that stage the embryo was about the size of one's little fingernail. All the RBCs growing in the liver made it easy to see as a little red spot in the embryo's abdomen. Each pregnant mouse had half a dozen embryos, so we could get enough cells for an experiment from one or two females. We suspended the fetal liver cells in tissue culture media and pipetted the cell suspension into microtiter plates with 96 wells that could accommodate small volumes (0.4 milliliters) of cell suspension. A milliliter is about 0.2 of a teaspoon, to give you an idea of what we're talking about. We added 0.1 milliliters of multiple dilutions of our erythropoietin samples, then added the cells and cultured the plates overnight in an incubator at 37°C. In the morning, to monitor for cellular proliferation, just like I learned to assay the proliferation of lymphocytes, we assessed the rate of DNA replication at the differing dilutions of erythropoietin samples. The DNA of a cell duplicates just before the cell divides, so the data consisted of counts per minute (CPM) of newly synthesized DNA at each dilution of the sample. On graph paper we plotted the CPM on the vertical axis and the dilution of the sample on the horizontal. By comparing the data from different erythropoietin samples, just like the interferon assay, we determined the concentration of erythropoietin of each sample. This was the first rapid, quantitative assay for erythropoietin ever created and led to our first paper together, in the journal *Experimental Hematology* (26).

Torg and I went on to collaborate with Edward De Maeyer and Jacqueline De Maeyer-Guignard, a Belgian couple, well-established in the interferon field at the Curie Institute. They met in Boston when both were postdocs with John Enders. Enders won the Nobel Prize in 1954 for the discovery that polio virus could be propagated in non-neurological cells, which led to the first polio vaccine developed by Jonas Salk from the University of Pittsburgh. Ed and Jacqueline were a real team of co-equals; they ran their lab and did experiments together. We all got to know each other and enjoyed *dejeuners* at a nearby bistro. They generously shared their knowledge of French cuisine, encouraging me to try new things like *au point Camembert*, a soft creamy cheese from Normandy, along with one more glass of red wine at the end of the meal — such a civilized custom.

The De Maeyers' postdoc, Larry Mobraaten, an American from the Jackson Labs in Bar Harbor, was busily purifying interferon — separating the protein molecule responsible for the interferon activity from all other proteins. It was an easy fit for us to work together. We decided to ask the question of whether interferon would inhibit the proliferation of RBC precursor cells in response to erythropoietin, just as I had found that it inhibited lymphocyte proliferation in the patients given polyIC at Dartmouth. Anemia (deficiency of RBCs) was one of the well-known complications of viral infections, but no one knew why a viral infection was associated with a fall in the RBC count. We postulated that anemia occurred especially during viral infections as compared with bacterial infections because viral-induced interferon inhibited the proliferation of the RBC precursors.

It turned out the answer was straight-forward. Larry's purified interferon preparations inhibited erythropoietin-induced fetal liver cell proliferation, which led to our second publication (27). Combined with my Dartmouth findings — that interferon inhibited normal lymphocyte proliferation — our results supported the growing idea that interferon inhibited the proliferation of *all* cell types, thereby contributing to the notion that a common mechanism underlies both normal and malignant cell proliferation. Obviously, I was pleased with this conclusion.

During the Christmas holidays, one of the postdocs in the lab offered to let us use his family's apartment in the French Alps for a week. It was in Serre Chevalier, a ski resort in southeastern France. I figured the Alps were always snow-covered, so the skiing would be great. We had all learned how to ski in the past year at Dartmouth, and we excitedly set out for our winter vacation two weeks before Christmas. The trip down through Burgundy was lovely, especially when we stopped overnight at a marvelous inn across from the Hospices de Beaune, site of the famous Burgundian wine auction that sets prices in November of each year.

When we arrived in Serre Chevalier, the snow was a no show. The village was practically deserted. Evidently, the season had not yet started. We learned that the snow usually accumulates over January and February, and spring skiing was the best in this area. But our spirits were still high. I figured we could cross-country ski on the patches of snow cover so far. No such luck. We discovered, upon awakening the first morning, that our car had disappeared. I went to the police station and realized I didn't know the French word for stolen. The police chief was able to decipher my heavily American-accented French and with many hand gestures took my statement. He admonished me that I should have had an *antivol* device. (*Voler* is the French verb for steal.) Our missing car was a used Fiat, vintage 1968, manufactured before they had *antivols*. I trudged back to the apartment. The good news was that they soon found the Fiat, out of gas and abandoned in a field on the outskirts of town. The presumption was that teens, probably drunk, took it for a joyride. The bad news was that the *voleurs* damaged the steering mechanism and the local garage had to send to Italy for the spare parts. We spent most of our holiday cooling our heels, waiting for the repair. I discovered that long soaking baths accompanied by a snifter of cognac soothed my anxiety. At the same time, it was great to be in the Alps and, when our car was out of the shop, to drive in the mountains. Looking over the drop-offs down into the abyss, Lynn was not as keen on the driving as I was. But I took it easy as we passed small crosses along the side of the road.

We arrived back in Paris the week before Christmas and planned to spend Christmas day with Torg and his wife and their five teenagers. They had rented a large Tudor-style house in Jouy-en-Josas, a suburb in southwest Paris not far from Versailles. For the holiday dinner, we bought a large *buche de Noel,* a traditional chocolate roll cake fashioned like a yule log. Because our minuscule French refrigerator could not accommodate it, we decided to leave it in our car, which was parked on the street just outside our building, figuring that it would stay cold enough overnight. Imagine my surprise when early Christmas morning I went out to check on the *buche* to discover that the Fiat was gone again. *Incroyable!* Torg drove into Paris to pick us up and we had a delightful Christmas dinner with all the trimmings at his house. Torg's teens took charge of our toddlers. The big kids entertained the little kids and vice versa. This was the first time we met Torg's wife, Barbara, a Scot whom he met as a Canadian tourist in Scotland when he was 19. Barb still had the quintessential Scottish accent and was fun to listen to. Two weeks later, the police found our car, abandoned in a field again. This time it remained in working condition, but the joyriders had devoured the *buche de Noel* and chocolate butter-cream frosting was smeared all over the interior. In the end, this Christmas was one for the memory books, and now we look back on that time as one when we snapped back, better than ever.

The Fredricksons and the Smiths became lifelong friends. Barb and Lynn explored Paris together, identifying their favorite restaurants. One in particular, Au Pied de Cochon (The Pig's Foot), was a famous eatery in Les Halles. Just prior to our arrival, the quartier's large market had moved to the outskirts of Paris, but the restaurant was still going strong, a favorite among the locals for its home-style cooking. Right nearby was a famous cookware store where Lynn started her collection of copper cookware just like Julia had.

In February we celebrated my 32nd birthday at a small restaurant not far from the Eiffel Tower. Little did we know it would lead to an escapade with a lasting effect on our family. There was a tremendous storm with freezing rain that night. Coming from New Hampshire, we weren't going to let a little rain and icy streets keep us home, so we ventured out. The restaurant, which had only a few tables, was run by a husband and wife; he was the chef, she was the hostess. Only one other table was occupied and the place was warm and cozy. We had a delicious dinner, sharing a red burgundy, '68. While the storm raged, we kept hearing a cat meow at the door adjacent to our table. Finally, our hostess let the cat in, which she explained was a local stray and frequent beggar at the restaurant. She gave her some cream, then let her out, but the cat again cried at the door. Lynn and I missed our two cats that we had to leave behind in Hanover, and when we left the restaurant, we looked for the stray cat. We found her crouched between two cars. After a chase, we caught her and took her home with us. The kids were ecstatic to have a new pet. The following week, Torg and Barb came to our place for dinner. Torg was a vet before he studied for his Ph.D. When he picked up our new pet, which we had named Chatte (French for a female cat), he suddenly said, "Are you aware that you have a pregnant cat?" That was news to us. Several weeks later we were the proud owners of three kittens, born hidden from view, under a bed. We found homes for two of them and elected to keep one of the males, which we named Tigre for his coat resembling a tiger's. When we returned from France, we had to smuggle both cats past the customs officers, which Lynn accomplished by sweet-talking and distracting them from opening the travel bags we put them in.

The experiments Torg and I did with interferon and erythroleukemia were innovative and productive for hematologists, but peripheral to my ambition of furthering my understanding of adjuvants and immunotherapy of leukemia. Two postdocs from Australia, Hermann Wagner and Marc Feldmann, had reported recently that it was possible to generate cytotoxic T lymphocytes in vitro by culturing T cells from unrelated individuals for a few days, then testing them for their capacity to kill in what was called a killer assay (28). I was keen to see if I could employ this killer assay to generate T cells that could kill leukemia cells. Torg was helpful because from his lab at Orsay he could deliver a ready supply of Friend and Rauscher virus-induced erythroleukemia cells for my experiments. Since I had never performed a killer assay, I set up some preliminary experiments to make sure that I could generate meaningful data before proceeding to determine whether I could make it work. When I mixed T cells from a mouse immunized with leukemia cells in a mixed tumor-lymphocyte culture, I succeeded in generating detectable killer, or cytotoxic, activity. Now I knew I could make it work in my own hands. It was important to me that I knew that the mixed tumor-lymphocyte culture method could generate cytolytic T cell activity. I planned to use this method to try to create a new way to treat leukemia.

Just about every weekend, Lynn, Kerry, Kelly and I explored a different quartier of Paris. We really needed to get our toddlers out of the apartment (as well as ourselves). We found out that the city's churches and parks had free concerts most weekends, and they became our focal point. One concert at Versailles featured four baroque trumpets in a small abbey and produced unbelievable sound that ricocheted off the stone walls. From there on after, the trumpet was one of our family's favorite musical instruments. In the summer, outdoor concerts in the parks provided the boys the space to run around and expend their excess energy before settling in to enjoy the music. One of our favorites was the 17th century Parc de Sceaux, a short ride from our apartment. The concerts on the lawns of this spacious park were also popular among the locals. In the colder months, we went with Torg and his family to hear Beethoven's *Fifth Symphony* with full orchestra in the Cathedrale de Saint Denis, which was quite impressive, and several concerts at Notre-Dame and Saint-Germain-des-Pres, the oldest church in Paris.

As we explored the countryside, Normandy, which was very close, just northwest of Paris, soon became our go-to province to get out of the city into the countryside. For our tour guide, we used Gourmet magazine's book, *Bouquet de France: An Epicurean Tour of the French Provinces*. A chapter on each province let us know what to see and where to stay and also had recipes for traditional provincial dishes. We learned about good French cuisine the old-fashioned way, by eating it at the source. In addition to the apples, superb butter, cream and cheese of Normandy, we discovered affordable, delightful country inns with unbelievable restaurants. Even though our sons were young, they quickly learned to appreciate French cuisine, developing discerning palates and quite good table manners. One weekend we visited Mont Saint-Michel, a medieval village atop a peninsula/island that juts out into the bay between Normandy and Brittany. When the tides change, the sea rises 45 feet between low and high tide very fast, changing the peninsula to an island. As I was unloading the luggage from our car, which I'd parked on the sand just below a dock, I didn't understand why people above me were screaming *"la mer, la mer"* and pointing to the sea. I then realized the tide was swiftly approaching our car. Just in time, I managed to toss the last suitcase onto the dock and escape to dry land.

Our year in Paris sped by. It was time to return home. Lynn had assiduously saved money for our return crossing. I went about making the arrangements for our departure. There was a big problem: how to part with our beloved '68 Fiat. I took it to several garages, and I'll never forget one old *garagiste* in his blue overalls who listened closely to the motor bending down into the hood for what seemed a long time, and then declared, *"Elle est morte."* Ultimately, time ran out and I abandoned it, leaving the car parked on the street because I couldn't find anyone to buy it.

After a delightful family vacation at the end of July in Etel, a tiny fishing village in Brittany, we sailed again on the SS France. We landed on August 8 in New York to be met with huge headlines, "NIXON RESIGNS." I took it as an omen for good things to come. While we were away, Ross McIntyre at Dartmouth reapplied for a grant and had just received word it would be funded. We quite happily resumed our New England lifestyle. Since we left dragging the flag as a result of Watergate, it was good to be waving it once again, especially after

We culminated our year in France with an idyllic vacation on the Brittany coast in a tiny fishing village, Etel.

having experienced Napoleonic bureaucracy vs. American efficiency. In retrospect, American efficiency is simply a different way of doing things, to which we were accustomed, while the French ways were foreign. The experience gave us a greater appreciation of the difficulties of immigrating to another land. We all matured in the process.

Having completed three years as a postdoc, I negotiated with Ross to join the faculty of Dartmouth Medical School as an assistant professor of medicine in the Division of Hematology. Lynn had given me an ultimatum that it was time for me to stop training and get a "real job." I agreed it was time. Even though I had not found a mentor who could prepare me to enter academia, I had seen how three different labs functioned and I learned several experimental techniques, such as the importance of quantification of interferon, erythropoietin and lymphocyte proliferation, as well as killer activity. I knew what I wanted to do and was confident I knew how to do it. I also wanted to be my own boss. I vowed never again to trust my fate to anyone else. If I couldn't get grant funding on my own, then I shouldn't be in the business. And science, for all it's pure advancement of knowledge, I realized was indeed a business. You are in business to create new knowledge that you must publish so you can use your documented progress to apply for new grants to create more new knowledge and stay in the game.

CHAPTER 4

The March to the Molecules

WE DROVE A RENTAL CAR from New York the five hours up to Hanover, a world away from the cacophony of Paris and our petite apartment. It was surreal to re-enter the Northeast Kingdom and the serene quiet of our spacious home again. Lynn, Kerry and Kelly were ecstatic to be back with their old friends. It was soon September and time for school. Kerry was six years old and ready for first grade. Kelly was four, starting preschool. It was amazing to watch and listen to the boys, who immediately dropped the sing-song cadence of speaking French for their resumption of speaking "American."

I was the ripe old age of 32. The very first thing that I did was sit down and start writing grant proposals as a principal investigator, or PI, the National Institutes of Health designation of the primary person responsible for the research proposed. I decided Dartmouth was the perfect testing ground for me. Other than Ross McIntyre, no one else there had immunological expertise, and Ross was established in hematology circles, not immunology. I was, in the words of Dartmouth's motto, *Vox clamantis in deserto* (A voice crying out from the wilderness). Setting the bar high, I knew I had to succeed on my own. Hanover, New Hampshire wasn't New York, New York, but if I could make it there, I could make it anywhere. As a postdoc, I had observed young people trying to become scientists without a clear vision of subjects on which they should base their careers. Like an adult version of follow-the-leader, they were choosing to work on the hot topics of the day. After learning a few techniques, they searched for projects on which they could apply their know-how. That might work for the short term, but scientific techniques are always changing. The trick is to learn how to be always learning. In his book, *Advice to a Young Scientist*, Sir Peter Medawar, who shared the Noble Prize with Macfarlane Burnet in 1960 and was knighted by Queen Elizabeth in 1965, propounded this logical requirement: "If you want to be successful in science, you should work on something important."

The questions then became what was important and who decided. I had a pretty good idea of the important work that needed to be done in the areas that interested me. From the experiments Torg and I did with erythropoietin, I knew it was necessary to purify the active molecule that promoted the proliferation of fetal liver cells. So one grant application I wrote was based on our erythropoietin investigation and focused on using the new rapid and quantitative assay that Torg and I created to identify and purify erythropoietin. Evidently, others deemed this goal a priority too because the National Heart, Lung and Blood Institute at the NIH had just issued a request for applications to purify erythropoietin from urine collected from individuals living at high elevations in the Andes. As a consequence of inadequate oxygen at that altitude, people were known to have elevated levels of erythropoietin in their blood and urine. I enlisted the assistance and expertise of one of the young biochemists at Dartmouth, and we wrote the application together as co-PIs. The biochemist would apply biochemical techniques to purify the molecule responsible for erythropoietin activity; I would test the samples and fractions for activity using the *in vitro* bioassay, which was quantitative and — thanks to a 24-hour turn-around — rapid. It was also capable of assaying the multiple samples we would generate with the biochemical separative procedures.

My second grant application centered on generating mouse T cells capable of killing leukemia cells. I decided this was an important question others hadn't explored. Taking a page from Professeur Mathé's approaches, I proposed experiments on a mouse system to provide "proof of concept" to justify similar approaches that could be applied to humans. Mathé had employed therapeutic vaccine injections of leukemia cells pooled from multiple third-party leukemic individuals. I proposed to duplicate this situation using mice, but with the caveat that it was possible to start with healthy, non-leukemic mice and therefore normal, healthy immune systems. With the advantage of an experimental system, it could be a prophylactic immunization, rather than therapeutic immunization that Mathé needed to use in humans. After the *in vivo* immunizations, I proposed to perform a mixed tumor-lymphocyte culture as an additional *in vitro* booster activation, or immunization, hoping to expand the killer cells before testing for killer T cell activity.

While I was setting up the lab that first fall, Ross McIntyre invited Robert Gallo from the NCI to visit Dartmouth. Bob was already a laboratory chief, the equivalent of a university department chairman, even though he was only a few years older than me. He was charismatic, persuasive, and well known. His lecture was in the amphitheater where we held medical grand rounds, and the place was packed. He told us about a new retrovirus they had discovered that he said was the first example of an RNA tumor virus isolated from human leukemia cells. He was convinced that human leukemia was caused by these viruses, primarily because retroviruses had been found to cause leukemia in several other species, including mice, rats, cats, dogs, cows, monkeys and apes. He emphasized that he was able to identify this new virus because his lab had found a way to culture human leukemia cells long-term, and these cultured cells produced a virus that was distinct from any known virus. Here's the ringer. The trick to culturing the leukemia cells was the discovery of a unique growth factor produced by a human embryo cell line. Bob's story sounded like science fiction to me, but extremely important, if true.

That evening, Ross and I took Bob out to dinner at the Hanover Inn where I told him I was trying to generate killer T cells that could kill mouse leukemia cells. I asked him if the leukemia patient from whom the cultured leukemia cells had been derived was still living and whether it would be possible to obtain their lymphocytes. He said yes to both. The patient was at the MD Anderson Cancer Center in Houston. Bob said he could arrange for blood to be sent to his lab at the NCI in Bethesda. I proposed to come to Bethesda to do experiments with the cultured leukemia cells and the patient's lymphocytes in a mixed tumor-lymphocyte culture. For me, this was an exciting, unique opportunity. I could bypass the mouse and go directly to the human to show tumor-specific immune reactivity. This would be akin to Mathé's pioneering bone marrow transplants on the Yugoslav physicians accidentally exposed to lethal radiation.

It was the spring of 1975 I loaded up my newly acquired little Fiat Spider with all the lab supplies and equipment I thought I'd need and headed south to Bethesda. I had an aunt and uncle who lived in Northwest D.C. where I could stay. Gallo's lab turned out to be quite unusual. At the time, NCI labs were undergoing renovation on the Bethesda campus so that the lab members were crowded together with contractors from a division of the government defense contractor, Litton Bionetics, in a nondescript one-story building in downtown Bethesda. Many of the larger labs at the NCI had contracting industrial labs at their disposal, thanks to President Richard Nixon's 1971 declared War on Cancer. The labs essentially doubled the number of scientists working for the lab chiefs. Compared to my nascent operation at Dartmouth, there was a huge number of individuals. The best part for the lab chiefs was that the tremendous expenditures they generated could be dealt with by the contractor's budget instead of the governmental NIH lab budget (29). Such a deal, eh?

Unfortunately, on a weekend not long before my arrival in Bethesda, an accidental freezer thaw ruined the vials of the frozen human embryo cell line. This was a disaster because these cells produced the growth factor that supported the leukemia cells. Without them, propagation of the new retrovirus was impossible, and I needed those leukemia cells for my experiments. The remaining original embryo cells that had not been frozen and were still in culture were producing far less growth factor than they had originally. Evidently, this was a common occurrence, and they would simply take another vial from the freezer to grow and produce the growth factor until these cells too stopped producing the activity. Without the frozen embryo cells, there could be no renewable source of growth factor for the leukemia cells and thus no retrovirus. However, when I got there, there were still some leukemia cells left growing in the dwindling growth factor supply from the original embryo cells, which I observed under the microscope, and they really looked like leukemia cells. Thus, I had no reason to doubt the veracity of Gallo's claims. Even so, Gallo's enemies took this situation as an opportunity to challenge his claim to have discovered a new leukemic growth factor as well as a new virus. When the patient's lymphocytes arrived from Houston, I was able to perform just one mixed tumor lymphocyte culture experiment before the supply of growth factor was exhausted and the leukemia cells ceased proliferating.

If the Bethesda foray was a bust for my experiments, nothing ventured, nothing gained. On the bright side, I met lots of people, most of whom were my contemporaries just beginning their careers. Two of them would become critical for my own future experiments at Dartmouth. Doris Morgan was a young postdoc who had obtained her Ph.D. in Houston working on culturing bone marrow and blood cells. She was an experimental hematologist. Gallo recruited her from MD Anderson to grow the leukemia cells and to generate retrovirus for the virologists in the lab. I also met Frank Ruscetti, a postdoc from the University of Pittsburgh whose Ph.D. was on bone marrow and blood cell cultures. With the demise of the embryo cell line in the freezer thaw, Gallo directed Doris and Frank to both begin searching for another source of the leukemia cell growth factor.

Doris had been a lab tech for several years before grad school. She was highly experienced in cell culturing and more skilled than other postdocs at this stage of her career. She was also meticulous and thorough, two characteristics that were almost prerequisites for this kind of cellular work. Frank, who grew up in South Boston in an Italian family, was gregarious and talkative with the classic accent of his native city. As I came to know him better, I would describe him as a cross between Rocky Marciano and Leonardo da Vinci, a curious mixture of a tough street fighter and a brilliant scholar.

Back at Dartmouth, I set up the experiments I learned in Mathé's lab, including the erythropoietin assay and the generation of killer T cells using mouse lymphocytes. A few months later, I received word that the erythropoietin grant application was rejected. Evidently, the review committee did not appreciate my new *in vitro* assay using mouse fetal liver cells. They required award recipients to use the conventional mouse *in vivo* bioassay in attempting to identify and purify the molecule. I had encountered for the first time the old boys' club that was our grant peer review process. I knew that the half-dozen established investigators who received the grants would never be successful using this assay. It took more than a week, and multiple samples could not be assayed easily using live mice. True to prediction, the erythropoietin molecule was never actually purified from natural sources. Eventually, the erythropoietin gene was isolated, which took another decade. I fired off a letter to the NIH officials informing them of their folly and felt a whole lot better. McIntyre told me that I was shooting myself in the foot.

They say depending on grants to fund science is like big game hunting: you eat what you kill. It might be more like wading through the tall grass, searching for prey. I stayed steamed up over the rebuffed erythropoietin application. Good news arrived, though, when my tumor immunology grant application *was* funded.

A year after returning from France, I found myself with plenty of money, comparatively speaking. I had been working with just one technician, Mary Ferm, a recent Phi Beta Kappa graduate of Oberlin College, my father's alma mater. She was great, enthusiastic and a quick study. Together we set up all the assays. I hired two more techs, both recent college graduates. Then Steve Gillis, a grad student in the Dartmouth Biology Department, joined us, and soon so did Paul Baker, a postdoc who had been a student of Torg's at the University of Connecticut.

With this fledgling team, experiments moved along well. We saw that we could generate mouse killer T cells *in vitro* and could increase the efficiency of leukemia cell killing by first immunizing the mice with leukemia cells from another mouse strain, just as Mathé had employed therapeutic injections of pooled leukemia cells from third-party individuals. These experiments provided our first data showing that a series of immunizations could generate tumor-specific killer T cells, and resulted in the first publication from our Dartmouth lab, which was accepted by the prestigious *Journal of Experimental Medicine* (30) founded in 1896 by William Welch, who also founded the Johns Hopkins School of Medicine. The journal was later taken over by the Rockefeller Institute in New York City.

One of the lessons of my postdoc fellowships was the importance of learning techniques directly from the experts who created them. The smallest details could make or break an experiment. Back from France, I contacted several NIH experts in monitoring RNA tumor viruses and visited their labs to observe and learn directly how to perform several assays. One was for an enzyme to identify retroviruses, another for the expression, or presence, of viral gene-encoded proteins on the cell surface and third for a cell culture assay for virus production.

In the following spring of 1976, I received a call from Frank Ruscetti. He told me that their lab may have accidentally come upon a way to keep T cells growing in culture long-term, apparently indefinitely. Here was a perfect example of serendipity in science. As the story goes, Doris Morgan had been charged by Gallo with looking for a leukemia cell growth factor by screening culture media derived from many different cellular sources. The lab had stored away huge quantities of media "conditioned" (influenced, modified) by PHA-stimulated lymphocytes. Peter Nowell from UPenn had reported in 1960 that PHA (phytohemagglutinin) is a plant molecule extracted from the kidney bean that initiated mitosis (2). His experiments suggested that the cells produced something that promoted cell growth. Doris placed human bone marrow and blood cells from leukemia patients into this medium because of recent reports of growth factors found in lymphocyte-conditioned media (31). Frank said that although Doris had failed to grow leukemia cells, she was successful in growing cells that looked under the microscope like lymphocytes, and preliminary tests showed that they might be T cells. However, the researchers could not be sure. Frank asked me whether I could perform our interferon assay on samples that he would send to me coded and blinded. It had been established more than a decade earlier that PHA-stimulated T cells produced interferon activities, so this experiment might support the idea that the cells Doris was growing were actually T cells (31, 32).

We tested the samples Frank sent for interferon activity and returned the results. Frank then broke the code, revealing which samples were positive for interferon. It turned out we were spot-on. We found interferon activity only in the samples from the cultured T cells that had been stimulated with PHA. Thinking over the implications of these results, I spoke to Frank again. I told him I thought their long-term cultured T cells could be important because although we had generated T cells that could kill leukemia cells, we could not keep them growing in culture for more than a few days. It would be great if we could grow them continuously and create large numbers of them. The T cells that Doris was able to grow did not perform any antigen-specific immunological functions, like killing, so it was not clear whether her cells were immature T cell precursors derived from the bone marrow or mature T cells capable of normal immunological functions, like killing. Ordinarily only immature cells are found in the bone marrow, while T cells capable of immunological functions had to have migrated from the bone marrow to the thymus before they could populate the blood and lymphatic vessels. One thing was clear. When Doris substituted fresh media for the lymphocyte-conditioned media, the cells immediately died; only the PHA-stimulated lymphocyte-conditioned media sustained the T cell growth. We presumed the lymphocyte-conditioned media contained something (activities, growth factors) that promoted the sustained T cell growth. In support of this idea, two separate groups had already reported a decade earlier in 1965 (33, 34) that lymphocytes produced mysterious mitogenic activities that promoted the proliferation of T cells in short-term cultures. Over the ensuing decade, 1965-75, more and more reports supported these initial findings for short-term growth. But no one had reported long-term T cell growth, beyond just a few days (35).

Most significant, no one had identified any molecules responsible for the mysterious activity. I proposed a collaboration with Frank to see whether we could use the lymphocyte-conditioned media to sustain the proliferation of our anti-leukemic killer T cells (36). Because all our killer T cell experiments were performed with mouse cells, we placed mouse cells into culture with mouse lymphocyte-conditioned media after both *in vivo* and *in vitro* leukemic cell immunizations, hoping that they would continue to divide.

We had a strategic lab meeting out on the front lawn of the med school to discuss our tactics. It so happened that this was one of those fabulous summer days in New Hampshire, just the right temperature and a crystal-clear blue sky. I put everyone in the lab — all the techs, as well as Paul, Steve and myself — onto this quest. It was a long shot and I did not hold much hope that this mysterious activity could support antigen-specific T cell growth. But if it worked, I knew that we would be onto something great. Immunological dogma had for years maintained that only antigens could activate T cell growth. Immunologists routinely placed lymphocytes into culture at a high density of a million cells per milliliter. Doris Morgan always used a tenfold lower cell density of 100,000 cells per milliliter. It turned out that made all the difference because later we found that the cells consumed whatever the growth factor was in the conditioned media; and when the cells grew to a high density, they rapidly ran out of growth factor and died. To our surprise and wonderment, we were able to sustain long-term growth of our T cells for weeks, and upon testing periodically, we found they continued to kill leukemia cells (37). I knew we were onto something big. Not only had we generated *antigen-specific* long-term killer T cell cultures, but we showed that *tumor antigen-specific* T cells could be generated. That meant that these cells might be useful therapeutically to treat and perhaps cure leukemia.

We sent our manuscript off to *Nature*, the renowned London-based scientific journal founded in 1869, fully expecting a quick review and publication because the results were so clear-cut, and I thought, consequential. I was stunned when we got a rejection letter almost as fast as the return post, which told me that they had not even sent it out for review. After I recovered from my initial shock, I got angry, which led me to be concerned that they just did not understand the significance of our unprecedented findings. These results had never before been reported. I wrote a one-page letter, explaining succinctly why this was such an important paper and asked them to send it out for review. They did, and the reviewers agreed with me. Our paper reporting the long-term growth of tumor-specific cytotoxic T cells was published in *Nature* on Bastille Day, July 14, 1977, the anniversary of the French Revolution. This I took to be a good omen, as I knew it was to herald a revolution for T cells and for immunology.

Eureka!

> The long-term culture of cytotoxic T cells overturned several dogmas and set the stage for a series of experiments that forged a new path in immunology and led to new therapeutic approaches. First, it had been accepted that proliferation of normal cells could never be extended beyond 50 cell divisions. This was thought to distinguish normal from malignant cells. It was clearly wrong as evidenced by our findings. A second dogma was that killer T cells could not proliferate indefinitely because the antigen stimulation had caused them to differentiate (that is. change into functional cells). The dogma said that differentiated cells could not proliferate. Wrong again. Our findings showed T cells that had differentiated to become killer cells still proliferated. Also, the differentiation of T cells to become capable of killing was thought dependent solely on antigen stimulation. Wrong once more. These revelations pointed the way to experiments establishing that the activity produced by the conditioned media, *not the antigen per se*, was responsible for stimulating the killer cells to proliferate. The genes coding for the molecules involved in the killer *mechanism* are now known to be stimulated by the growth-promoting activity, not antigen. It was only possible to discover this because we had abandoned the notion that antigens are responsible for everything, all by themselves.

No sooner than we had this new awareness and a few experiments under our belt, I got another call from Frank. He told me Gallo did not want his lab to work on the T cells anymore. Frank said they could no longer collaborate with me because Gallo wanted to try to reproduce their experiments using a leukemia-specific growth factor to culture human acute leukemia cells. He needed these cells to look for their production of human leukemia viruses. Frank said Gallo was turning the whole T cell project over to Ron Herberman, an NCI immunologist. I thought that he should have turned it over to my lab since I was clearly interested, but Frank said that Gallo wanted it to remain at the NCI. He reassigned Doris to generate PHA-stimulated lymphocyte media for Herberman's lab to try to repeat the long-term T cell growth in his lab. Frank was reassigned to search for other cell sources of the factor that promoted the leukemia cell growth. I have found that my brain will do all the heavy lifting without any conscious effort on my behalf. After I have slept on an issue, it is always clear what I should do about it. With the clarity of 24 hours to mull over this development, I decided that this was a good thing. Now we could focus on our own experiments and not worry we would be competing with Gallo's lab, which was huge compared with my fledgling group, sort of like David and Goliath.

As soon as we had grown our long-term cytotoxic T lymphocyte lines (CTLL) for a few weeks, it was obvious we had generated a never-ending need to produce conditioned media to keep our cells growing. The two words, factor and activity, essentially are both meant to convey the meaning of a *substance* of unknown composition that appears to promote a biological response. Logic dictated that it must have some kind of molecular structure. So our top priority was to identify exactly what was responsible for promoting growth. To do so required constructing an assay to determine how much growth factor was contained in each batch of medium. I cannot emphasize enough how important our assay for T cell growth factor was for everything that came after. We published the details of the assay in 1978. By 2016 it would become the second most cited paper in *The Journal of Immunology* in the 100 years since its first issue in 1916. Others had devised assays for activities found in lymphocyte-conditioned media, but they revealed only a yes or no answer as to whether there was detectable activity in a batch of medium. They were not quantitative; that is, they did not determine how much factor was present in a batch. They showed only if it was there — or not. Obviously, if one could determine the *amount* of factor in one batch vs. another, it would be possible to determine an amount or concentration that would consistently result in continuous cell proliferation. The assay itself was patterned after the interferon assay I learned at Dartmouth and the erythropoietin assay that Torg Fredrickson and I devised in Villejuif (26). These experiments underscored another of Peter Medawar's points in *Advice to a Young Scientist*. It is possible to gain a great deal of information and understanding by simply quantifying phenomena that others had observed but did not fully understand (38). The ability to quantify, or measure, the amount of growth factor activity was huge in the annals of immunology. It ushered in the first step towards a molecular definition of the regulation of immune responses. With this paper, we coined the term *T cell growth factor*.

Gallo's decision not to pursue long-term T cell growth left the field wide open. We were one of the very few labs in the world that initiated work in this nascent area, and we were left unfettered for nearly the next five years. Here again, it embodied the Dartmouth motto, Crying out in the wilderness. Had we been at a larger center, word would have gotten out on the grapevine and attracted many more competitors. It was only after five years when we had defined how important TCGF was that Gallo woke up to it and would claim that *he* "discovered" TCGF. In fact, his group at the NCI, especially Doris Morgan, had discovered something important — that it was possible to culture T cells longer than a few days using lymphocyte-conditioned media — but the NCI group had not identified and purified the molecule in the conditioned media. Without the molecule, it was impossible to proceed further experimentally. With our new TCGF assay, we knew that we could identify and purify the responsible molecule.

Since Gallo had turned the project over to Ron Herberman, I decided to invite him to visit Dartmouth to find out what he was doing. Ron was a mild-mannered but crafty guy. Earlier, I had visited his lab at the NCI to learn the killing assay, and I saw he held his cards close to his chest. At Dartmouth, he gave a seminar that was unrevealing, but his visit was memorable because of our conversation afterwards at lunch at the Hanover Inn. Whatever would we have done without the Hanover Inn? He told us that his lab was having difficulty repeating Morgan's work and couldn't get human T cells to grow longer than a week or two. We told him that we were not surprised at his lack of success. We too had had more success with our mouse cultures than with maintaining human cells in culture. We were pleased (secretly) he was having difficulties reproducing Morgan's findings. But here's the clincher: Herberman asked us whether we had cloned our long-term T cells. We told him no, we had not, but that one question was worth my invitation to visit us.

Several technical advances in the 1970s allowed immunology, if not all biomedical research, to progress from the cellular to the molecular level. One was the creation of monoclonal antibodies, reported in 1975 (39). Because monoclonal antibodies attached only to individual molecules and could be obtained in large quantities, this technology could be used to identify and purify actual molecules that previously had been identified only as ephemeral factors or activities. Another advance, which we first reported in 1979, was the creation of monoclonal killer T cells (40). A clone is a population of cells that are all descendants of a single cell. If it could be proved that a population of cells was actually derived from only one T cell, that is, monoclonal, then heretofore impossible experiments could be done. It would be truly ground-breaking. Paul Baker volunteered to attack this experimental problem in our lab. Paul's experiments cloning our cells worked immediately. The cloning efficiency was phenomenal, approaching 100 percent (40). Previously, cloning of malignant cells without a special growth factor usually was 2-3 percent efficient.

The immunology community was ecstatic over our methods to create T cell clones, which were quite simple, provided you had the proper amount of TCGF. Everyone wanted to create their own

T cell clones that reacted solely with their favorite antigen. Our methods were rapidly adopted worldwide and became standard for creating monoclonal T cells. These cells were critical because with unlimited amounts of cloned cells, biochemical and molecular experiments could be done to determine how the cells functioned; for example, how they recognized only one specific antigen and how they killed other cells. That was impossible with limited amounts of heterogeneous cell populations. Another reason relates to the unique nature of the immune system. Each person has half a trillion T cells, but each T cell is capable of recognizing only one specific antigen. It was impossible to begin to identify the molecule responsible for the antigen specificity of a single cell using a T cell population comprised of so many different T cells.

Eureka!

> Monoclonal antigen-specific T cells permitted experiments that allowed us and others to delve into the molecular workings of the T cell for the first time. The molecular nature of the T cell antigen receptor was revealed. Also, now decipherable were the molecules responsible for the ability of T cells to kill other cells, as well as the molecules responsible for T cells to help bone marrow-derived B cells differentiate into antibody-forming cells. In essence, monoclonal T cells enabled the transition from cellular to molecular immunology.

While my work was progressing smoothly, we had a family situation that needed fixing. Kerry was not faring well in first grade in the Lebanon school district. Lynn, a former first grade teacher, visited the school a few times and concluded that our son's teacher definitely left a lot to be desired. Kerry was a hyperactive child used to continuous physical activity. His teacher wanted him to sit still all day at his desk. Fortunately, we had an option to enroll him on a tuition basis in the Hanover elementary school, which used an open classroom style. Kerry recovered quickly and did remarkably well. Yet this was only a stopgap solution because we could not afford to pay tuition indefinitely going forward. Lynn started house-hunting in Hanover and found a great Victorian built in 1893. We moved in during the summer of 1976 and I knew that we had found our home at last. It was on a one-way street two blocks from Hanover's Main Street and overlooked Mink Brook, a tributary of the Connecticut River. We were a mile from my lab at the med school, so I walked to work in the winter and in the warm months rode my French Motobecane bicycle. The kids in the neighborhood spent many hours playing down in the ravine. Kerry and Kelly learned fly fishing, first in the brook, then the river, which was just a few miles downstream and about a half mile wide at that point. While the boys were fly fishing, I spent many hours writing outside in the garden during the summer. When the cold weather came, I retreated to my cedar-lined study with a small fireplace, stoking the fire while I wrote and read journals. Looking south from our front porch, we saw just forest with Mount Ascutney in Vermont in the distance, a single prominent peak, sort of like Mount Fuji. A Dartmouth botany professor who lived in the house for many years had gathered wildflowers from all over New England and planted them in the garden. Probably our home's best feature was the large kitchen with a big bay window overlooking the wildflowers. Lynn was ecstatic.

Lynn decided not to resume her teaching career, at least not elementary school, now that she had two active boys of her own. Instead, she began the Parisian Kitchen, a cooking school, where she taught courses in French cuisine. I encouraged her to pursue this endeavor. She was a natural teacher and comfortable being the center of attention, the one with the expertise. Over the course of the next six years, she acquired a following of folks in and around Hanover who wanted to learn how to cook like Julia Child. The boys and I benefited immensely as she tried out each new dish on us and perfected her repertoire. The kids turned into expert culinary critics as a result.

Lynn, teaching her art in her school, The Parisian Kitchen.

In the lab for me at the end of the '70s, the revelation that TCGF, instead of the antigen per se, was responsible for the proliferation of the clones of antigen-selected cells was the bright light that made me understand for the first time how the immune system works. As hypothesized in 1957 by Macfarlane Burnet, who incidentally was knighted by the Queen in 1978, the clonal selection theory posits that the antigen's role in immunity is to pick out, or select, the individual cells reactive with the antigen among the vast numbers of cells in the body, each destined to react with myriad other antigens. When I realized that TCGF then drives the massive expansion of the numbers of clonally selected cells, I suddenly understood that after an antigen selects which cells will participate, TCGF becomes the key regulatory factor in the response by the immune system. I knew this should be our focus going forward. We would be able to manipulate the immune response, by manipulating the availability of TCGF.

Eureka!

> The notion that there existed an internal control of the T cell immune response was an epiphany. As a physician I had never understood why immunologists considered the immune system as special, distinct from every other organ system. It was thought to be regulated entirely by external antigens that entered the body from the environment. Yet in the context of evolution and natural selection, it seemed to me that the immune system should be regulated internally, like every other system in the body. The understanding that the control of the immune response is internally regulated opened a vista I had not seen before in immunology. If you stop to think about it, when life made the transition from a single cell to multicellular organisms, mechanisms had to be devised for an organism's multiple cells to communicate with one another. That brought me to the conclusion that molecules produced by the cells must have been the way that communication between all cells was possible.

One extrapolation of this concept was that the expansion of a population of antigen-selected T cell clones should depend on the availability of T cell growth factor. Following discussions with the team, Paul Baker volunteered to perform a series of experiments to test this hypothesis. Accordingly, after addition of the antigen to the wells of the microtiter plates to initiate selection of antigen-responsive cells, any TCGF produced was removed by replacing the medium with fresh medium daily. Alternatively, TCGF was added daily to try to ensure a maximal concentration, which we could determine as a plateau of the rate of proliferation. Adding more TCGF could not extend the proliferative rate beyond a certain point. The results supported our hypothesis, which I viewed as fundamental to the concept that TCGF was central to the regulation of the immune response. Without TCGF there was little proliferation and without cellular proliferation there could be no immune response. We prepared a manuscript entitled "The effect of T cell growth factor on the generation of cytotoxic T cells" and sent it to the *Journal of Experimental Medicine*, thinking that this journal among all others would understand and appreciate our findings since I considered it the most sophisticated of the journals that published immunological reports.

No such luck. It was getting to be a familiar story. We soon got our manuscript returned — rejected. To add insult to injury, this journal had a policy of not sharing its critique of rejected submissions. I could not imagine why they would reject such an obviously (to me) important paper. So I phoned the editor who signed the letter to find out what the problem was. I appreciated Henry Kunkle taking my call, but he said that he did not remember the details. He checked his notes from the editorial meeting and said that the editors who reviewed what we'd sent them thought it was just another of those "factor" papers and was not very important. I was apoplectic and blurted out, "They did not understand because this was *not* just another 'factor,' as our findings were central to all of immunology." Henry, who was one of the senior immunologists, got flustered and clearly wanted to end the conversation. I had never met him, but I knew that he was old school, a gentleman professor, very measured and low key. I also knew that he was powerful in immunology and that if he wanted to, he could torpedo my chances of ever having another paper accepted by the *J Exp Med.* Before he could get off the phone, I volunteered to come down to New York to give him a seminar on why our paper was so important. He tersely declined. Ultimately, we published the paper in *The Journal of Immunology* (41), and a few months later Henry invited me down to give a seminar at the Rockefeller University. I had a great visit with him and knew he now understood the significance of our work.

My first grant had been awarded for three years rather than five because I was a new investigator. This meant that I had to start writing for a renewal of the funding after only one year as the review process took a year. We had made remarkable progress. I was not worried about the renewal. Still, the peer review of my application came back with the recommendation *not* to fund. One of the reviewers commented in the summary critique that "evidently Dr. Smith thinks that it is possible to grow antigen-specific normal T cells indefinitely," implying that what we said we'd accomplished was not true. I was shocked. After I recovered, I decided to fly to Bethesda to talk to the people in charge of my grant. I took what remained of my grant monies and bought a plane ticket to Washington. I boarded a 20-seater on Air New England (Dartmouth students dubbed it "Air No Engine") from Lebanon to Boston's Logan Airport, then on to Washington National. It was to become a familiar route.

When I met with my program officers, I told them what we had achieved thus far. I also said I knew there were naysayers about our long-term cytotoxic T cell lines. I could see one of the problems was that I was new to the game, and I was coming from Dartmouth, a place most people in immunology had never heard of. There were essentially no card-carrying immunologists at Dartmouth in the cold Northeast Kingdom. I was so frustrated. I asked them, "How do you play this game, anyway?" Their response was basically: "Dr. Smith, this is serious business, not a game." I went back home without any idea how I was going to move forward. Lo and behold, a few weeks later I heard that they had reversed the decision of the study section and they were extending my original grant for two years. This meant that I would have to begin writing another renewal application in a year, but the reprieve gave the naysayers time to become more aware of who I was and what we had accomplished.

Using our T cell growth factor assay, we determined that both T cells and macrophages were necessary to produce the growth activity, but we could not say definitively which cell type was responsible for actually producing it. This became important because another T cell growth-promoting activity, termed lymphocyte activating factor (LAF), had already been reported to be a product of macrophages when stimulated by bacterial lipopolysaccharide (LPS), which is a potent immunological adjuvant (23, 24). From my experience with the adjuvants polyIC and polyAU and their role in of enhancing T cell growth via their effects on macrophages, this made perfect sense to me. Accordingly, we prepared LAF according to the recipes in the literature, but when we tested it using our assay, we found no TCGF activity. After that we took to saying, "LAF is a joke" in that it clearly did not make our T cells proliferate. We did find that LAF augmented TCGF production by purified T cells (42). Further experiments in collaboration with Joost Oppenheim and Lawrence Lachman, experts in macrophages and LAF, confirmed our initial results (43). The 1979 Second International Lymphokine Workshop was held that year in Ermatingen, Switzerland on the shores of Lake Constance that borders Germany, Switzerland and Austria. After we presented our results, many people were quite excited, and we discussed their implications late into the night in the bar. Our discussions led to the proposal to rename these two activities, LAF would become interleukin-1 and TCGF would become interleukin-2, since they appeared to work in series, one after the other. We created the word interleukin to designate that the activities carried messages between *(inter)* white blood cells, or leukocytes, thus *leukin*. Several investigators forcibly reasoned that if there were two of these activities that could be identified and separated based on their functional attributes, there would be more, so a numerical nomenclature made sense. Now at the time of this writing, 43 interleukins have been discovered. In fact, the naming of interleukin-1 and interleukin-2 was premature because no one had purified either activity or shown that a single molecule was responsible for either one. It turned out that the creation of this new nomenclature was problematic for me, in that it implied that interleukin-1 was discovered before interleukin-2 and was the first interleukin. In the end, we purified TCGF first, but when I told my sons that I discovered the first interleukin molecule, their reaction was, "Sure, Dad, that's why it's number two!"

Within just a few years of returning from France, I was on my way to understanding the molecular basis of how T cell growth occurs and discovering a new and unprecedented molecular "hormone-like" system that regulates the immune response. Moreover, since immunological adjuvants like polyIC and polyAU, as well as bacterial LPS (lipopolysaccharide) and bacillus Calmette-Guerin, were all known to be potent macrophage stimulants, for the first time we were peering into the black box that regulates immune responses and coming up with the molecular understanding of the mechanisms responsible. All these adjuvants augment interleukin-1 production by the macs, which in turn augments interleukin-2 production by the T cells, so that these two interleukins function as a bimodal amplification system that rapidly and markedly ramps up T cell expansion and the capacity of T cell populations to respond.

With a little smile, I said to myself, "See, Professeur Mathé, it *is* possible for such a young, inexperienced investigator like me to understand how adjuvants boost immune reactivity. One must just do reductionist science, progressing from the cells, to the activities, to the molecules that give the signals to the cells regulating their behavior." It is like descending from 30,000 feet to the ground and then entering the Star Wars bar scene, suddenly seeing all the funny looking creatures.

What allowed me to have this vision? For starters, my background in medicine with instruction in the disciplines of endocrinology and pharmacology was foundational. Endocrinology is a branch of physiology and medicine that studies glands that secrete hormone molecules into the blood, which then circulate to affect different cells and tissues. Pharmacology is the study of drugs. Hormones and most drugs interact with specific receptor molecules on target cells and tissues. I now realized the immune system also uses hormone molecules and receptors. Without the perspective of how other systems in the body are regulated that I obtained from medicine, I never would have come to this eureka moment. My serendipitous route through the basics of lymphocytes along with the brief forays into interferon and erythropoietin had prepared my mind to think about how these activities might function at the molecular level. It was Pasteur who said, "Luck favors the prepared mind."

By 1980 I had undergone a metamorphosis from a young physician hoping to succeed in science to a man obsessed, turned on to science around the clock. I knew that what we had already uncovered would transform the science of immunology. The way forward for the interleukin revolution was clear. It would be a revolution because these new immunological activities would become defined as hormone molecules and would provide the focus for therapies previously unknown in immunology and medicine.

CHAPTER 5

Immunological Messenger Molecules Are Hormones

Lynn transformed this 1826 colonial house into her classic French restaurant.

Lynn's life also changed dramatically during the 1980s when she decided that she had taken her cooking school as far as she could in our small town. She needed a bigger challenge. A chance opened up when one of her bachelor uncles passed away and left her a small inheritance. She used it to start a new French restaurant in Hanover, Café La Fraise (the Strawberry Café). She purchased an 1826 Colonial house — a classic New England two-story, wood-frame, yellow saltbox, the last remaining single-family home in the business district. She found a young and skilled construction company, Domus, to transform it into a restaurant. She designed the kitchen and dining rooms and made one of the front rooms into a food shop where she sold French bread, deserts, pastries and home-made country pates, all prepared in her kitchen. She patterned the bar on the second floor after similar establishments we got to know in France. After six years of teaching cooking, Lynn knew what she was doing culinarywise. She had developed her expertise. Café La Fraise became a popular meeting place for an afternoon snack for the boys and their friends, who quickly renamed it Café La Friends-Eat-Free.

It changed my life too. I was at first a little worried that she could bring off such an ambitious project. It's one thing to have a successful dinner party for six, quite another to run a busy restaurant. But she was a natural as a chef and a manager. I welcomed and encouraged this new endeavor. We had been growing apart due to the daily pressures of family and my career. Many people at this stage in life have a hard time finding a reasonable work-life balance. Lynn had had enough of my tales of the great hotels and restaurants I visited on my frequent trips. These days when she tells the story of our lives back then, she always says, "I was left at home with the shit and string beans, and he was staying in five-star hotels." And she blossomed. She maintained that as a mother of two boys, the challenge of directing her restaurant was a walk in the park. She later told me she knew then that she could become independent. She could go her own way, if necessary, because she had found a way to support herself.

Years afterward as I looked back and tried to understand and explain how our marriage survived, it boiled down to three key words: harmony, synchrony and compromise. There cannot be a lead singer and a backup in a marriage that lasts. These two singers, to achieve true harmony, must have equal roles. It has to be a duet. When Lynn started Café La Fraise and became even busier than

I was, we were in harmony with equally responsible roles. And we finally achieved synchrony as our life rhythms merged. Asynchrony in a marriage due to differing pressures of work and family responsibilities often leads to irrevocable breaks. Lynn inserted the last key word, compromise, pointing out that we could have never been harmonious or synchronous if we both had not learned to compromise.

The restaurant was halfway between our home and the lab. I stopped there on my way to work. We would share an espresso and a croissant hot from the oven and chat quickly about our plans for the day. In the evening, I would stop by just before service, then go home to the kids. After the last order had been filled, around nine o'clock, Lynn called me at home and asked what I would like for dinner. I could have anything from the menu or the special of the day. With just the last guests still in the dining room, we discussed our day and tried to help each other solve our respective management issues. We were totally together. For our relationship, it was heaven on earth. We realized that postdocs, lab techs and students were similar to sous chefs, cooks, dishwashers and servers. They all were motivated to do a good job, but most times needed to be told exactly what to do and how to do it. As Lynn said many times, she would check with them to make sure that they were doing their work correctly. She held them accountable. What a concept? Truth be told, when I started my own lab, I had an expectation that everyone should know what to do and how to do it. Lynn taught me otherwise.

Kelly was 13 and Kerry 15 in 1983 when Lynn started the restaurant. Both boys were consumed by soccer and hockey. Their hockey team played at Dartmouth's Davis Rink. Sounds great, but it was an outdoor rink with just a roof and open sides. It got really cold in the deep winter, like 20° below zero. I thought that my poor toes would drop off after each game. Lynn would come from the restaurant, whenever possible, to cheer her sons on. Kerry developed into a skilled goalie. Kelly was a terror on the ice, although he spent a lot of his time in the penalty box. With school and sports, both boys were fully occupied, leaving Lynn and me to our all-consuming work. What self-respecting teenage boy wants his mother or father to know what they're doing all the time? For Kerry and Kelly, with access to all of Dartmouth's facilities, it was like growing up in college. Kerry took an interest in gymnastics and worked out after school with the Dartmouth team whose members were keen to offer him advice and help.

In the lab, we were treading on new territory. By 1980 we had uncovered the first data suggesting that a previously unrecognized molecular regulatory system regulated the immune system. It was essentially an immune hormonal system, or an immunoendocrine system, much like the nervous system is a neuroendocrine system that functions using neurohormones that are called neurotransmitters. I realized the next few years of experiments were going to be important for establishing new paradigms that, hopefully, would lead to a novel understanding of how the entire immune system was organized and regulated. I was confident this awareness and understanding would lead to the discovery of new therapies.

The '80s proved to be a golden age for interleukins. Other scientists identified additional interleukin molecules, discovered the genes encoding them and, through genetic engineering (cutting and pasting the genes into bacteria) huge quantities of these newly discovered gene products were made available for study. More people were attracted to the field, and it became a new international discipline.

It was clear from the outset of the decade that if we thought the immune system was akin to a hormonal system, the first question was: What constitutes a hormone? First used at the beginning of the 20th century (44), the word is derived from Greek, meaning to set in motion or to excite or arouse. Classic hormones, such as adrenalin, insulin and glucocorticoids, came to be known as molecules released into the blood, usually from distinct glands, that circulate to stimulate distant target tissues or cells. Hormone molecules interact with receptors, or receiver molecules, on the surface or inside the cells of the target tissues. It is now recognized that receptors are responsible for both the *specificity* and high *affinity* (that is, strength) of hormone binding to target tissues that allow the hormone effect to occur at very low concentrations of the hormone molecules. In addition, by definition, a hormone not only binds to a receptor; a true hormone receptor, as opposed to just a binding site, must also transmit information to the target cells, which elicits a characteristic biochemical response that changes the behavior of the target tissue. This is now referred to as *signal transduction*, meaning the transfer of a physical force of binding into a biochemical reaction or vice versa. If we thought that T cell growth factor was actually a hormone, then our top priority was to identify a TCGF receptor in the quest to understand how the hormone promoted the proliferation of T cells. At this point, even though the interleukin terminology had been coined, I decided to stick with TCGF until an actual molecule was identified.

One other unique characteristic of the immune system set it apart from the more classic endocrine system, where a distinct gland produces the hormone molecules, secreting them into the blood. The source of TCGF was not a gland, but rather T cells that were mobile, circulating throughout the body in the blood and lymphatic vessels (a system parallel to arteries and veins). Also, since the target cells, the T cells themselves, were also constantly moving throughout the body, the immune hormonal system was novel, akin to a mobile army with its own communication system, like mobile phones.

Regarding hormones, it's worth noting again how Louis Pasteur's maxim, "Luck favors the prepared mind," enters the story. Not long after I came back to Dartmouth from France, but way before we discovered TCGF, I had become acquainted with another faculty member, Allan Munck, whose lab was one floor up from mine. About 15 years older than me, Allan was a full professor of physiology and an expert on hormone receptors. In the late '60s, he had described the first evidence that glucocorticoid hormones caused physiological responses by interacting with specific receptors that he found to be present in all cells and tissues he tested. Glucocorticoids are perhaps best known today as drugs used to suppress immune or inflammatory reactions and to treat leukemia and lymphoma. Some of the most commonly prescribed are prednisone, hydrocortisone and dexamethasone. These are all classified as steroid hormones produced from the fat molecule cholesterol by the adrenal gland situated on top of the kidney. The term steroid is from the chemical description of molecules that consist of four rings of five-carbon atoms bonded together. Other steroid hormones include the female sex steroids, like estrogen and progesterone, and the male sex steroid, testosterone. They all differ by just a few atoms. A deficiency of glucocorticoid hormones is called Addison's disease, which most famously afflicted President John Kennedy. Over-production of glucocorticoids is known as Cushing's disease, named after Harvey Cushing, the American neurosurgeon who in 1932 first reported a patient with a pituitary tumor at the base of the brain, which secreted hormones that caused the adrenal glands to produce excess corticosteroid hormones. There are two adrenal corticosteroids. Glucocorticoids regulate the biochemical reactions that burn our food, like the sugar glucose to produce energy, and mineralocorticoids that regulate the levels of minerals, such as salt, in the body. Receptors for both corticosteroids exist in every cell of the body and these hormones and receptors are vital for life.

Allan had used rat thymocytes (immature T cells) to develop his assay for glucocorticoid receptors because the thymus in mice and rats undergoes a rapid involution (shrinkage) when they receive a large dose of glucocorticoids (45). The thymus is a sensitive glucocorticoid hormone target tissue. In the mid-'70s, Allan and I started to collaborate to detect glucocorticoid receptors in human lymphocytes. Allan was originally trained as a chemical engineer and was so mathematically oriented that he actually thought in equations rather than sentences. He became interested in biology and hormones as a postdoc at MIT in a lab focused on steroid hormones. He had used the principles of radiolabeled hormone binding established in endocrinology to demonstrate glucocorticoid receptors for the first time in 1968 (45). When Allan and I initially discussed studying the effects of glucocorticoids on mature human T cells isolated from blood, the idea to measure the number of glucocorticoid receptors was the furthest thing from my mind. Allan persuaded me that this should first be accomplished in order to establish data indicating that mature T cells just like thymocytes had the capacity to respond to glucocorticoid hormones. I soon found myself immersed in the intricacies of detecting and counting the number of hormone receptors per cell.

In the life-long learning process that is science, it helps immensely to have a good teacher. Allan was just that. He became the mentor I had been seeking. Our collaboration, which resulted in my first *Nature* paper (46), showed that the number of glucocorticoid receptors on normal human T cells increases when the cells are activated by a mitogenic substance, such as PHA, that stimulates cells to proliferate and undergo mitosis. For several years, Allan and I worked together monitoring glucocorticoid receptors and studying their effects on both normal and malignant lymphocytes. I was interested in the glucocorticoids from my medical training at Ohio State, Yale and the NCI, all of which taught me they had dramatic suppressive effects on lymphocytes. I recalled the devastating effects of the high doses of glucocorticoid prednisone on the young girl with leukemia I treated in Baltimore. But the molecular mechanism for these effects remained obscure. This question made steroid hormones and their receptors and their suppression of T cell proliferation another target of my research.

We knew from Peter Nowell's experiments of the 1960s that glucocorticoids prevent the T cell proliferative response stimulated by PHA, but we didn't know how they accomplished this feat (47). Once we had established our bioassay for T cell growth factor, it was obvious that we could examine how glucocorticoids suppress lymphocyte proliferation after activation with PHA. We figured that if TCGF was the real mechanism responsible for T cell proliferation upon antigen or mitogen stimulation, and not antigens or mitogens themselves, then corticosteroids might block the production of TCGF, or block its action, or both. Given our TCGF assay, it was easy to design experiments that would let us know.

The first results of our experiments were clear-cut. Glucocorticoids completely prevented the production of TCGF, and also inhibited its action by 25-30 percent, thereby completely suppressing lymphocyte proliferation This provided an explanation for the potency of glucocorticoids on the immune system when used as therapeutics (48, 49). Our experiments established a paradigm or model for examining the mechanisms of several additional immunosuppressive drugs that soon became available in the early '80s, and which revolutionized organ transplantation, (50, 51). Among them were cyclosporin-A, tacrolimus and rapamycin. Like glucocorticoids, both cyclosporin-A and tacrolimus were found to block the production of TCGF, while rapamycin was eventually found to inhibit its action (51). These drugs synergize and complement each other, so that they are more efficient and less toxic in preventing transplant rejection than when used individually. Prior to the publication of our data in 1979, kidneys were the only organs that could be successfully transplanted. For example, the first heart transplant, in 1967, made the South African surgeon Christiaan Barnard world famous. Still, this first attempt and then many others, including lung, liver and pancreas transplants, although possible surgically, all failed because the recipient's immune system rejected the transplanted organs. The new immunosuppressive drugs that impaired the production and action of TCGF essentially put the surgeons back into the operating rooms, making the suppression of transplant rejection more effective, thereby creating a whole new field — transplant medicine.

Eureka!

Our experiments illustrated the power of understanding the molecular mechanisms responsible for T cell proliferation. It was like coming into the light from utter darkness. This encouraged us to think that it might be possible to uncover all the intricate internal workings of lymphocytes. In addition, these experiments for the first time showed that a classic endocrine (hormonal) system, first discovered by John Langley in the 19th century, exerts a regulatory control over the immune system, discovered almost a century later. We realized that the immune system is an immunoendocrine system, integrated into other molecular regulatory controls that the body exerts to ensure a proper living balance. The body is a finely tuned machine that consists of thousands of molecules that interact with one another and change each other in the process. Seems logical now, but back then this notion was heretical, especially for immunologists. These data were the first to show physicians that the immune response is regulated by molecules produced internally, and therefore one could hope to manipulate the system therapeutically by inhibiting the production or function of these internal molecules. Alternatively, we began to think about using these molecules to enhance or stimulate the immune system in situations where there was an immune deficiency.

Prior to this revelation, immunologists considered the immune system to be quiescent until an antigen was introduced, activating and expanding clones of antigen-reactive cells, which somehow facilitated removal of the antigen, followed by a return to a quiescent, resting, state. Of course, this is what happens, but a hormonal immune system constitutes the molecular mechanism whereby these functions are accomplished. If antigen alone had been responsible for activation of the immune system, the only means to manipulate immunity would be to manipulate the antigen, either by increasing it or decreasing it. This would be way too complicated because there are so many different antigens, like half a trillion. That would make it impossible to manipulate therapeutically. The epiphany that immune hormones exist and actually are responsible for the onset, magnitude and duration of immunity was astonishing. It opened a whole new therapeutic world. Transplant rejection, allergy (non-microbial antigens) and autoimmunity (self-antigens) are reactions that became amenable to therapeutic immunosuppression by blocking the production or action of these immune hormones.

Because of my training in medicine, I had been schooled in the concept that hormone molecules and many drugs interact with their target tissues by binding to specific receptor molecules. Had I only received graduate training in immunology, it would not have seemed perfectly logical that immunological regulatory molecules would work like classic hormones and therefore one might be able to manipulate the immune system *therapeutically* by altering these hormones and their receptors. This new awareness that endogenous, or internally derived, molecules regulate the immune response and that this was not due solely to antigens external to the body from the environment as professed by the immunological dogma provided both the impetus and the direction for our research for the next several years.

Although we had only a partial picture of how the immune system is regulated, I knew that we were onto the key to unlock the secrets. I was fully prepared by 1980 to embark on the next set of experiments. Recall that when researching erythropoietin with Torg Frederickson in France in 1973 and 1974, we had discussed how red blood cell growth-promoting erythropoietin exerts its effects on the target cells. I had spent hours in the institute library in Villejuif, studying the literature trying to find out as much as I could about hormones and, in particular, their receptors. Steroid hormones, including glucocorticoids, and the sex hormones — testosterone, estrogen and progesterone — all interact with receptors located in the cytoplasm of the cell. Steroid hormones being relatively small molecules (molecular size ~ 350 atoms) synthesized from cholesterol, pass right through cell membranes to find their receptors in the cytoplasm. By comparison, larger protein hormones like insulin (molecular size ~ 5,800 atoms) interact with receptors embedded in the cell membrane. In the early '70s, besides glucocorticoid and estrogen receptors, scientists had identified only three other receptors—for insulin, epidermal growth factor and nerve growth factor. The only way then to assay for the presence of receptors was via a hormone binding assay: The hormone was tagged with a radioactive atom (a radioisotope), then incubated with the cells in small plastic tubes. After a brief interval to establish a steady state of the hormones interacting with the receptors, the cells were separated from the unbound hormone in a high-speed centrifuge. Then a radioactive counter gave two sets of data: CPM bound to the cells and CPM that remained in the supernatant, the liquid media outside the cells. So, it was a simple assay.

The sticky wicket to doing these hormone-binding assays was obtaining pure hormone that was radioisotope-labeled. In the case of glucocorticoids, Alan could just buy them from a biological supply company because they were small molecules and had been, synthesized and purified in the 1950s. But we could not simply buy radiolabeled TCGF. It was too new and no one had purified it to homogeneity so that it could be made free from contamination by other molecules. If we had tagged an impure mixture of molecules with a radioisotope, we could never have used it for a binding assay. Fortunately, we had progressed in this regard to make our own pure TCGF because our bioassay was well-suited to measuring the TCGF concentrations in multiple samples. Plus, and this was critical, we could show that there were no other molecules present (52). We radiolabeled the TCGF molecules by producing TCGF in media containing radiolabeled amino acids that were incorporated into newly synthesized TCGF molecules, just like we used radiolabeled nucleotides to monitor newly synthesized DNA molecules (53). Once again, recent advances in physics and chemistry made pioneering work in biology possible. Without radioisotopes, which began to be used in the 1950s after discoveries in nuclear physics, biological research could never had progressed to the molecular level. With radioisotopes available it was possible to tag molecules so that they could be detected. Otherwise, there was no way to know whether a molecule was there or not. They were just too darned small — a thousand-fold to a million-fold smaller than a cell, which could be seen using a microscope.

Allan helped us develop the radiolabeled TCGF-binding assay, the first such receptor assay created for these newly found molecules. His sage advice ensured no repeat of the errors of early endocrinologists. Our radiolabeled TCGF-binding assay became the prototype for all cytokine receptor assays (53). For example, Allan insisted we perform the assay at 37°C, noting that the radio-labeled insulin binding assay's developers performed it at 4°C, just above freezing. This introduced an artifact, something observed in an experiment that is introduced as a result of the procedure used for the observation. Molecules slow down their movements in the cold, so that there was no way to be sure of measuring *true* hormone receptors, that is, those responsible for the biological or organic, effects of life. At freezing cold temperatures, molecules nonspecifically bind or cling to each other. Consequently, it was impossible to relate the hormone binding data at 4°C to the biological activity of the hormone, which obviously only occurred at normal body temperature.

As the data on radiolabeled TCGF binding accumulated, it appeared the binding sites obeyed what endocrinologists had established as *true*, or real, hormone receptor characteristics. A true receptor must have a binding site for a hormone, but it also transmits a signal to the cell. According to Allan, true hormones and receptors display the following characteristics:

1. There are always a limited number of binding sites per cell, so that it is possible to bind all binding sites, or *saturate* them, simply by increasing the hormone concentration.
2. The binding can only be detected on hormone-responsive cells.
3. The binding is specific; no other hormones or factors compete for the binding.
4. The hormone concentrations that stimulate the cell coincide with those that promote radiolabeled hormone binding.

With the data obtained by the experiments Allan helped us design, I ran up to his lab two steps at a time and excitedly said, "So we're dealing with a real bona fide hormone and receptor, correct?" Allan thought for a while, then very professorially answered, "No, not until you can demonstrate negative feedback regulation. You see, all hormones must have a way to regulate themselves." Otherwise, he explained, once triggered, or turned on, there would be no way to turn the system off. And since TCGF causes the cells to proliferate, that might result in never-ending proliferation, the hallmark of malignancy, cancer.

That took the wind out of my sails, but it also gave me an idea of how malignancy might develop.

We discussed how we might test for negative feedback regulation of TCGF production or activity and performed several experiments along this line. We were not confident we could demonstrate negative feedback. But with the only assays we had available — the TCGF bioassay and the radiolabeled TCGF binding assay — we recognized it was not yet technically feasible. We knew we needed more information about TCGF and its receptor, and we needed to be able to track more immediate effects of TCGF binding to its receptor in additional ways. We could only monitor cellular proliferation via DNA synthesis, which took 12-24 hours to begin after TCGF bound to its receptors. We realized that we had to understand the events that occur inside the cells more rapidly upon TCGF receptor binding — within seconds, minutes and hours. This realization would thrust us to ponder exactly how the physical binding of TCGF to its receptor molecule on the outside of the cell could become transduced into biochemical signals inside the cell that ultimately led to DNA replication and cell division. And even though I had no immediate idea as to how to approach this question, I knew that we would be entering a whole new world of signal transduction.

The radiolabeled-TCGF binding assay allowed us for the first time to calculate the concentration of the TCGF molecules necessary for TCGF to bind to the cells. We already knew this equaled the TCGF activity that promoted T cell proliferation we had measured in our TCGF bioassay. But we had no idea how the TCGF activity related to the concentration of actual protein molecules. We were shocked to find that this concentration was exceedingly small. Knowing the size of the molecule, we could calculate for the first time the amount of TCGF protein detected by both the radiolabeled TCGF binding assay and the TCGF bioassay. By this time, not a surprise, the number was also infinitesimal. This told us that to end up with enough pure TCGF protein to be able to progress to the next steps of experiments, both on the nature of the TCGF molecules as well as experiments to further understand the TCGF biological activity, we would need to accumulate at least 1,000 liters of conditioned media and use several biochemical steps where we would lose a large proportion of the TCGF protein at each step. This understanding was eye-opening and a game changer as it presented an insurmountable task.

Luck would have it that a cell line established from a patient with T cell leukemia was found to produce TCGF (54, 55). We cloned cells from this cell line and selected a very high TCGF producer. Repetitive bioassays yielded TCGF concentrations one hundred-fold in excess of the concentrations that normal T cells produced. This helped immensely because it reduced the TCGF-conditioned media that we needed to begin our biochemical purification procedures from 1,000 liters to only 10. This was certainly feasible for our small lab, so that we were definitely in business.

I now knew that this immune hormone system rested on a firm molecular and experimental foundation.

CHAPTER 6

Receptors on T cells

WITHIN A DECADE AFTER ITS introduction to immunology in 1972 by Leonard Herzenberg of Stanford University (56), the flow cytometer had replaced the microscope as the tool of choice to study immunological phenomena. Flow cytometry allowed exquisite sensitivity to not only detect, but also to count molecules on the surface of individual cells. Here we go again, getting quantitative. Beginning in the 1960s, the earliest instrument for that purpose was a large box that took up almost a whole room and required a physicist on duty to keep it running. It consisted of a contraption that sucked cells in a suspension down a narrow tube so that they passed by a laser beam as single cells. Laser light is familiar to all us now as the device in stores that reads bar codes instantaneously. The word laser is an acronym for "light amplification by stimulated emission of radiation." Laser light consists of a single color or wavelength. In the lab, if a molecule can absorb light of a short wavelength and emit light of a longer wavelength, it is said to fluoresce. If such a molecule is attached to a monoclonal antibody, the laser can impart increased energy to the fluorescent-labeled antibody, which makes it detectable by the instrument. By the time Doreen Cantrell and I started to use the flow cytometer at Dartmouth, it had been reduced to the size of a refrigerator. However, there was only one instrument at Dartmouth. It cost more than $100,000 and required a full-time technician to operate it. Combined with monoclonal antibodies, which can bind specifically to individual molecules, the flow cytometer transformed immunology from a descriptive science using a microscope and a hand-clicker to count cells, to a laser-computer capable of counting the number of molecules on the surface of individual cells, and eventually to detecting and counting molecules inside cells. Since molecules are a thousand-fold to a million-fold smaller than cells, it's easy to understand that this was a significant advance. It again illustrates how dependent molecular work in biology depends on detection methods that physics and chemistry made possible.

Doreen joined my lab as a postdoc in the summer of 1982. She had just completed her graduate studies in the UK, where she focused on flow cytometry to analyze lymphocyte surface molecules. Her arrival was opportune because in collaborative experiments with another NCI investigator, Thomas Waldmann, we had just identified a monoclonal antibody called Tac that was reactive with the TCGF receptor (57). This antibody had been generated by a postdoc in Tom's lab, Takashi Uchiyama from Kyoto University, whom I had met on the first of my many visits to Japan. Takashi informed me that the antibody was known as Tac not because it meant "activated T cells," which everyone assumed, but in fact, he said, it was named after him, Takashi, "so that it should have been spelled Tak!" With the flow cytometer and the new Tac monoclonal antibody, we could now observe and quantify the TCGF receptors on individual cells. The next series of experiments nicely illustrated the technological revolution that the flow cytometer and monoclonal antibodies brought to immunology.

In the fall of 1981, I had taken a three-month mini sabbatical to visit a new friend, Max Schreier, at the Basel Institute for Immunology. Together we searched for evidence of a B cell growth factor. Max had developed helper T cell clones, and we surmised that they would produce a B cell growth factor as a mechanism of their "help". Thus, it was in Basel where I interviewed Doreen as a potential postdoc. She had answered an ad that I placed in *Science*. I asked her to come over to Switzerland from England so that I could interview her. Compared with the doctoral training system in the States, which usually extends for five to seven years following four years as a college undergrad, the English and European systems are fast-tracked. Students can enter the university at the age of 18 and finish six years later with a doctorate, either a Ph.D. or M.D. While I thought Doreen was young for a postdoc when we met, I was impressed by her resume, especially by the fact that she had taken a year off after her high school to be an au pair with a family in Thailand before beginning the university. It just seemed like a gutsy thing to do. I liked Doreen immediately. We spent most of the day in the institute cafeteria drinking coffee and talking science. Because of her British accent, it was easy to fall into the trap that we Americans often feel towards Brits — that they sound naturally smarter than us — and Doreen talked very fast. I could hardly get a word in. Initially, I was concerned that she had done her Ph.D. studies on natural killer cells (NK cells), which at the time I viewed as being suspect because they were not specific like T cells; that is, they could kill any cell recognized as nonself and they were only a minor subset (about 10 percent) of lymphocytes circulating in the blood. I drilled her nonstop with question after question, trying to understand how she approached and solved scientific problems, especially whether she thought NK cells were a real phenomenon and why. As the morning turned into the afternoon and then evening, I knew she was perfect for the lab. She was experienced in the burning questions of immunology, such as the nature of the T cell receptors for antigens, and the mechanisms by which helper T cells helped B cells make antibodies and killer cells killed. She was quick, analytical, and, of utmost importance, ambitious. When the subject of leaving England for the U.S. arose, she seemed to think that the main problem was her cat, which she would have to leave behind. That clinched it for me. I figured that she would do just fine. We agreed that she would come to Dartmouth the following summer.

I met Doreen at the airport in Lebanon, got her situated at the Hanover Inn and suggested we go to Peter Christian's, an English-style pub owned by a good friend. I figured that might help her deal with the culture shock of New England vs. Olde England. We talked science until the bar closed. Even though Doreen had just flown from England, she did not seem jet-lagged, and she seemed just as excited as I was with all the new experiments to do. We were eager to embark on the next series of experiments characterizing how the TCFG receptor activates T cell proliferation.

Doreen immediately went to work with the flow cytometer. When we got our first data showing the distribution of TCGF receptors on a population of activated T cells, it immediately struck us that there was a marked difference in the number of those receptors expressed by individual cells within a population: some cells had a high density of TCGF receptors, others had much lower densities. We were astonished that there was a thousand-fold difference. This was a revelation because with our radiolabeled TCGF binding assay, the data told us only the average number of receptors on the total population of cells in the test tubes. Now we wondered whether a high or a low density of TCGF receptors made any difference in how a cell responded.

Of course, it did.

Eureka!

Doreen Cantrell's meticulous experiments showed that the cells with the highest density of TCGF receptors start to replicate their DNA and divide sooner than cells with fewer receptors (58). Sounds simple now, but back then it was a huge breakthrough. We surmised there must be a certain number of receptors that had to be triggered by TCGF before the cell could make the irrevocable (all-or-none = quantal) decision to replicate its DNA and then to divide. If the cell only had a low density of receptors, it would take longer to achieve this critical number of triggered receptors. Eventually, all the cells within a population would replicate their DNA and divide, but it took some cells much longer than others. I thought of it as the New York City Marathon; after the gun goes off, everyone starts to run and the racers spread out with the fast runners crossing the finish line before the slower ones. If this were true, there must be some way the cell could actually count the number of triggered receptors. These considerations set the stage for the next several years. Our description of these experiments and findings was published as a major article in *Science* entitled "The Interleukin-2 T Cell System: A New Cell Growth Model." Looking back, I see that it represented the culmination of all our hard work defining TCGF and its receptors at the molecular level. It clearly was one of the best papers, if not the best, to come out of my lab (58).

As detailed in the introductory paragraphs of this paper, the unexplained questions we answered regarding variable cell growth characteristics went back 50 years to the 1930s and involved all living cells that had been studied, including bacteria, yeast, and avian and mammalian cells. The critical new knowledge that made it possible was the capacity to attribute the growth characteristics of the cells to the critical molecules, TCGF and its receptors, which actually directed the cells to grow. I began to realize my goal of understanding how the growth of normal cells is regulated. This new cell growth model system would provide a basis for our future experiments in New Hampshire to find the answer to cancer.

A postdoc from Sweden joined us soon after Doreen arrived. Martin Gullberg grew up in Umea, Sweden, near the Arctic Circle. He was the star student of Eva-Lotta Larsson, who together with Antonio Coutinho had also realized early on that both lymphocyte activating factor and T cell growth factor constituted a hormone-like bimodal amplifying system that promotes T cell proliferation (59). Eva-Lotta had invited me to be the opponent for Martin's dissertation, a unique custom for the Swedish Ph.D.. Universities in Sweden invited an expert in the field to come to the university and essentially play devil's advocate in an unrehearsed presentation before the entire university community. It was a big deal. We had a lively debate. Martin was great. He did not cave into any of my challenges to his findings or their interpretations, and the audience gained an appreciation of the significance of his hard work. I guess this was a test for me as much as a test for Martin because Eva-Lotta contacted me later on to see if Martin could join my lab as a postdoc.

Martin and Doreen hit it off and became inseparable in the lab, continuously discussing their experiments. Even as they grew to be best friends, they were quite competitive. Doreen said Martin was dogmatic and stubborn and often frustrated her, but they always came to some sort of compromise, or truce. Doreen discovered that the word "please" did not exist in the Swedish language; she felt that explained a lot. Despite these stylistic differences, we all had a good time together and frequently found ourselves at the Hanover Inn bar gossiping about the science and scientists of the day. Another of our oases was the Dartmouth Faculty Club, a short walk from the lab. Situated in a picturesque old Colonial house, we would repair to one of the upstairs conference rooms for a *dejeuner a la Professeur Mathé* for our weekly lab meetings, usually on Friday afternoons. When the discussion regarding our experiments became tense, Martin would reach for his Swedish snuff box, which always made me smile.

Around this time, Maurice Landy introduced himself to me at one of my presentations at a meeting. Maurice was then an editor with Academic Press, but prior to that he had a long career as an immunologist with the National Institute of Allergy and Infectious Diseases, having begun his studies in the 1950s. He was the father, or perhaps grandfather, of our research field, and one of the pioneers focused on soluble factors discovered in lymphocyte culture media. I certainly knew who he was. He came up to me after my talk and said, "I can help you, Kendall." That got my attention. "Really, how can you help me?" "By editing your papers," he answered, then asked me to send him a draft of my next paper when I had it in a reasonably final form. I did that and got it back right away by express mail (remember, this was before email) with red marks all over it. He didn't change what I said, just how I said it. He transformed the report of my research into a literary piece. Isn't it marvelous how people seem to appear in your life just when they're needed? Over the next five years, I sent Maurice all my papers and even my grant applications. He might change my adjectives or adverbs or rearrange my sentences, placing the end at the beginning or vice versa. And for sure, he got rid of dangling modifiers. This was the beginning for me of learning how much fun writing could be. Maurice stressed that "a scientific report must be concise, tightly integrated and persuasive". If he told me that once, he told it to me a thousand times. He reminded me that the first readers would be the reviewers. Therefore, one must write to persuade the reviewers first and foremost.

With Maurice's instructions in mind, I would write in the morning, then go on to other endeavors. Even though I might have stopped writing, I would think about what I had written throughout the day. When I reread the paragraphs the next day, I found myself rearranging the sentences or rewriting them, reshaping each paragraph to make it as 'concise, tightly integrated and persuasive' as possible before moving on to the next paragraph. Now, rereading my papers from the 1980s, I am proud of how well some of the most important ones turned out. For me now, writing is a creative endeavor, just as enjoyable and rewarding as designing experiments or giving a talk. All these skills must be practiced and mastered in order to become a successful scientist.

Towards the end of the decade, I became acquainted with a young scientist working at the Dana-Farber Cancer Institute in

Boston. I met Ellis Reinherz when we were placed on the same symposium program at an immunology meeting. I happened to speak first, then took a seat and listened to what Ellis had to say. His talk was mesmerizing. He explained in superb detail how he had one monoclonal antibody that recognized all human T cells, another that recognized about 60 percent of human T cells in the blood and provided "helper" activity for immune responses, while another monoclonal antibody recognized the remainder of human T cells that had "killer" activity. Sitting in the audience, listening to Ellis, I said to myself in a cartoon bubble, "This guy is really smart, maybe even smarter than me. I'm not going to compete with him. I'm going to make him my friend!" Which is what I did. That evening we went out to dinner, beginning a friendship that still is going strong today. After I got to know him better, I decided he was not that smart after all. He later told me he was "educable," which I found amusing. He was also suffering from inbreeding, so characteristic of prestigious educational institutions. After graduating from prep school near Boston, he went to college and med school at Harvard, did his residency at Massachusetts General Hospital, and was a hematology fellow at the Dana-Farber Cancer Institute in Boston. I figured with that degree of inbreeding I should take pity on him. I concluded he probably had supernumerary nipples.

Doreen was well along in her experiments looking into the nature of the TCGF receptor when we drove down to Boston to visit Ellis at Dana-Farber and present a seminar on our recent findings. There we met all the members of his lab, including his chief tech, Rebecca Hussey, who was gifted at growing T cells. Also present was Stuart Schlossman, who had been Ellis's postdoc mentor in immunology at the Dana-Farber. Doreen presented our experiments showing that the TCGF receptors were detectable only on the surface of antigen-activated T cells. She stressed the point that TCGF via its receptors, not any antigen, determined the growth response of T cells initiated by antigens. During the discussion after the talk,

Stu Schlossman asked her how she could be confident that the Tac monoclonal antibody recognized the TCGF receptor molecule. We both felt this was obvious from our data showing that the Tac antibody competed with TCGF for binding to the receptor. It also blocked T cell proliferation even though an antigen was still present. You see, until this point all the people interested in the nature of the T cell receptors for antigens monitored the proliferation of the cells following the addition of an antigen. They assumed that the antigen was responsible for the proliferation that they monitored. We couldn't figure out where Stu was coming from. I thought maybe he had not been paying attention. Doreen and I rolled our eyes at his suggestion that the Tac antibody reacted with something trivial. It became evident during an intense discussion between the various lab members that Ellis's lab was working on another T cell surface molecule involved somehow in T cell proliferation, but after much contentious discussion, we weren't exactly clear what they had.

On the ride home to Hanover, Doreen and I discussed what we both thought was a bizarre exchange with the Boston types. It dawned on us they probably had discovered the T cell antigen receptor but did not want to tell us because their findings were not yet published. Sometimes in science, you can't trust even your best friends. To be fair to Ellis, we still had not become close friends, so his caution was understandable, given the huge importance of the discovery of the molecular nature of the T cell antigen receptor (TCR).

We were soon able to read all about it. In February 1983, Ellis's group reported the first definitive evidence of the molecular nature of the T cell receptor (60). Over the next few months, a series of reports (I don't know how he had time to write them) showed the global immunology community that the T cell antigen receptor has multiple components: the signaling portion of the receptor, which they called CD3 (cluster of differentiation-3); plus the chains that act as accessory receptors, CD4 and CD8; and third, the newly coined alpha and beta chains, which are the proteins that actually bind to the antigen (61, 62). These experiments were definitive because Ellis used TCGF-dependent T cell clones of specific antigen reactivity. This enabled him to show that his monoclonal antibodies would block only the antigen stimulation of the appropriate clone of T cells and not clones with reactivity to other antigens.

For example, one investigator in particular, James Allison, had reported the generation of a monoclonal antibody reactive with mouse lymphoma cells in 1982 (63). He was attempting to raise antibodies reactive with a presumed tumor-specific antigen. He did generate a tumor-specific antibody and speculated it might be reactive with the T cell antigen receptor. But not knowing the antigen-specificity of the lymphoma's T cell receptor, he had no means to test for that possibility. Other investigators, including a team from the University of Colorado, headed by Philippa Marrak and John Kappler, would report subsequently that they had isolated a monoclonal antibody reactive with surface molecules of a mouse T cell hybridoma (64), which blocked antigen-stimulated TCGF production. They also found as did Ellis, that the cell surface molecules were comprised of two chains that were the antigen-recognition components (alpha/beta) of the T cell antigen receptor complex. However, they had no data on the other chains making up the TCR complex because the mouse CD3 signaling molecules and the mouse CD4 chain of the accessory TCR complex were yet to be discovered for several years. Thus, Ellis Reinherz's report was first, complete, and correct, on the molecular nature of entire TCR complex.

Just a few months later, in the summer of 1983, the 4th International Congress of Immunology was held in Kyoto, Japan. Ellis and I traveled together and were roommates during the Congress. At the arrivals terminal in Tokyo, as we waited in line to clear customs, an immunologist from Columbia University College of Physicians and Surgeons jumped the line and attacked Ellis verbally for his

Journal of Experimental Medicine report, questioning whether he had really discovered the T cell antigen receptor. Ellis was very cool and did not respond to this attacker. Clearly, this was going to be a riveting meeting. That evening, before the Congress had officially started, Mark Davis wowed an impromptu audience in a special brief presentation on how he and Stephen Hedrick, as two young postdocs at the NIH, had identified and cloned a complementary DNA (cDNA, a copy of the mRNA) that they speculated encoded, or contained, the DNA code for one of the antigen-binding chains of the mouse T cell receptor (65). Their findings were reported simultaneously the following year, 1984, in *Nature* together with a group from Toronto led by Tak Mak, who identified a human receptor counterpart cDNA (66). Rumor had it that Mak had cloned several cDNAs from a library, but he did not realize that the human sequence was similar to the Davis-Hedrick mouse cDNA. Following the meeting, Mak went fishing in his library of thousands of cDNA molecules again and found the human counterpart to the mouse cDNA. However, neither of these groups could be sure that they had in fact identified the T cell antigen receptor chain cDNA because they had no way to test for antigen specificity, which could only occur at the protein level. That these cDNA molecules in fact encoded the information for the amino acid sequence of a chain of the TCR and not some other protein was only confirmed by comparison with the Dana-Farber group's amino acid sequence of the human beta chain protein that they had proven had antigen specificity using their T cell clones (67).

Ellis and I spent a great deal of time talking science over our Japanese breakfasts of miso soup and rice, and over Japanese whisky late at night. I told him that all our data indicated that the T cell receptor promoted T cell growth by activating the production of TCGF and placement of its receptors on the T cell surface.

That meant that TCGF was responsible for triggering T cell growth, not the T cell receptor per se, or all by itself. This was the issue that Stu Schlossman had had difficulty accepting at Doreen's seminar. At breakfast one morning, I proposed to Ellis and he agreed that we should collaborate to settle this point since we, among all others in the world, had the unique cells and molecules necessary. When we returned from Japan, Doreen, together with postdoc Stefan Meuer and Rebecca Hussey, the chief tech from Ellis's lab, performed a series of experiments that left no room for the naysayers on exactly how the T cell immune response occurs (68): Most T cells do not have detectable TCGF receptors on the cell surface. Upon antigen (or monoclonal antibody) triggering of the T cell receptor complex, the TCGF receptors rapidly appear. If TCGF is present, after a lag of about 24 hours, DNA replication ensues, followed by cell division. However, if TCGF is blocked by monoclonal antibodies reactive with TCGF itself or its receptor, growth is prevented, even though the T cell antigen receptor has been triggered. Our two labs had now become the epicenters of research and discovery for both the TCR and TCGF. These data set a new paradigm, and became a whole new way of understanding how the immune response occurs, based for the first time on the newly discovered hormone-like molecules of the immune system.

Eureka!

> This new molecular awareness sounds so simple and straightforward now. Back then, it was revolutionary. The immunology cognoscenti had for decades believed that *solely* antigens triggered T cell growth. They were correct. However, there is a molecular mechanism that the T cell antigen receptor employs to accomplish this task — activating the synthesis and placement of TCGF receptors on the cell surface and the synthesis and secretion of TCGF into the surrounding tissues by the T cells themselves. This revelation meant that the immune system is regulated by an internal hormonal system, just like all organ systems in mul-

ticellular organisms. The timing, magnitude and duration of the T cell immune response are all ultimately dependent upon and thus regulated by TCGF via its receptors (69). In this instance, the term hormone — from the Greek, *horman*, meaning "to set in motion, to urge on" — connotes that signaling molecules interact with specific cellular receptor molecules to cause their effects. The importance of these concepts for therapeutic manipulation of the T cell immune system will become evident as our story progresses.

CHAPTER 7

Molecular Structure Emerges as the Key to Recognition

LEST WE THOUGHT THAT WE knew everything there was to know about T cell growth, one of the original perplexing observations stemming from our first experiments came when we placed mouse spleen cells into culture. As Ellis Reinherz originally reported, in the blood about 60 percent of T cells expressed the molecule characteristic of helper T cells, CD4, while about 40 percent expressed the CD8 molecule characteristic of cytotoxic T cells. However, after adding TCGF-conditioned media and letting the cells proliferate for a week or two, the vast majority (90 percent) expressed only the CD8 molecule. Doreen's subsequent experiments placing a similar mixture of these two kinds of human lymphocytes into T cell growth factor also found that the vast majority of the cultured T cells expressed the CD8 molecule. When Martin joined us, he chose to look at this phenomenon more closely. First, he replicated Doreen's findings, then set up a series of experiments comparing the growth of the two different cell subsets, CD4 and CD8. Every day he would remove a sample of the T cells to count and identify the surface molecules (CD4 or CD8) using the flow cytometer together with monoclonal antibodies reactive with each cell surface molecule. He found that the CD4+ cells (that is, those with cell surface CD4 molecules), ceased proliferating after one week of growth, even if he replenished the supply of TCGF, while the CD8+ cells kept growing for several more days. After 12-13 days, the numbers were up to 10 times higher for CD8 than CD4. Despite our confidence that we would now be able to figure out why this was happening based on the quantitative data of the two different T cells, we ended up only with the interpretation that somehow the CD4 cells but not CD8 cells became refractory, that is, resistant, to the growth-promoting effects of TCGF (70). It would be 20 years before this enigma in the complex regulation of the T cell immune response could be explained, but in retrospect, these were probably the first data suggesting that TCGF-induced negative feedback is central to a regulatory network dampening T cell proliferation, just as Allan Munck had predicted "for a true hormone-receptor mechanism you must demonstrate negative feedback regulation."

Before we had purified whatever molecule was responsible for scoring positive in our TCGF assay, we knew it remained possible that more than one molecule comprised the growth-promoting activity. If that were the case, it would be impossible to identify and purify just one molecule. Soon after we developed the TCGF radiolabeled binding assay that enabled us to identify and characterize the receptors, we knew we had to purify the TCGF molecule to homogeneity. This feat would vault us ahead of the rapidly crowding field. By 1981 we were no longer unknown in the immunological world; many labs were trying to do exactly what we were doing. The bottleneck to moving forward, especially to determine both the biological and biochemical characteristics of TCGF and its receptor, meant that we needed to prove that our TCGF preparations were pure, that is, homogeneous, uncontaminated by other molecules. We had made considerable progress in this regard using analytical biochemical methods, so we knew that a single protein was responsible for the biological activity, and we knew the molecular size (~15,000 atoms), but we still could not generate enough pure TCGF protein to proceed to the next level of experiments. That would require large amounts of pure protein, in the milligram range (a milligram is one thousandth of a gram = one small feather), not the microgram range (one millionth of a gram). We needed to generate a monoclonal antibody reactive with TCGF because then there were methods to use the antibody to rapidly purify in a single step the TCGF molecules from any other molecules that might be present. This would allow us to end up with larger quantities of pure molecules, in the milligram range, a thousand-fold larger than the microgram range we'd achieved with the available biochemical methods. The problem with the biochemical methods was the significant loss of activity and protein with each separative procedure. After five or six steps, we had pure TCGF activity but minuscule amounts of TCGF protein. This underscores the huge difference between the activity, which could be detected in the nanogram (10^{-9} gram, one billionth) protein range vs. the one million-fold larger milligram (10^{-3} gram, one thousandth) ranges needed for the next steps.

We had to show purity. It was the only way forward.

My friend Max Schrier was at the Basel institute for Immunology in Switzerland where George Kohler had been a graduate student. As a postdoc in Cesar Milstein's lab at the University of Cambridge, George had generated the first monoclonal antibodies in 1975 (39), so that the Basel Institute and Max were well versed in the art of making monoclonal antibodies. When I told Max what we were planning, he said, "Good luck, but I don't think you have a prayer to be successful." At that time, no one had made monoclonal antibodies reactive to a small protein molecule of only 15,000 atoms like TCGF. Most of the monoclonal antibodies created thus far bound to much bigger things, like cells, which are a thousand-fold to a million-fold bigger than molecules, and thus had many more antigenic sites for antibodies to recognize. Max told me that making monoclonals was akin to a craft, and that it took a long time to master it. I took his skepticism as a challenge.

Around this time, I hired a new tech, Maggie Favata, who was originally from the U.K. Maggie took on several ongoing assays and she was adept at cell culture, which was just what we needed at this juncture. In order to create monoclonal antibodies, mice were immunized with a specific antigen. This immunization resulted in the expansion of lymphocytes that had differentiated into antibody-producing plasma cells. The mice were then sacrificed, and their spleen cells were combined with a cell line derived from malignant plasma cells, termed plasmacytoma cells. By adding agents that fuse membranes, the normal and malignant plasma cells were melded together. The result was a cell hybrid, termed a hybridoma, that had the continuous growth characteristics of the malignant cell but still secreted the antibody specific to the immunized normal plasma cells. Then the task was identifying the antibody-secreting hybridomas reactive with TCGF among all the hybridoma cells secreting antibodies that reacted with other antigens. This was a crucial step, and one where we had an advantage because we had developed the defining bioassay and the TCGF receptor assay. We could use one assay to confirm the other. Which is to say, we had it all covered.

The first order of business was to generate a large amount of lymphocyte-conditioned media to produce and purify enough TCGF for immunizing several mice. Because the immune response is dose-dependent, we figured we would need at least 30 micrograms per immunization, which we calculated would require at least 30 liters (about seven gallons) of conditioned media to inoculate the mice with initial and booster doses. Maggie and I set to work to generate that amount. We needed a source of large numbers of normal human lymphocytes. We usually just drew blood from one another and separated the lymphocytes from the red blood cells by centrifugation. This could generate only about 100 milliliters (one tenth of a liter), so I contacted one of the otolaryngologists, the ear, nose and throat specialists, at Mary Hitchcock Hospital, who could provide us with tonsils removed in routine tonsillectomies. Tonsils are lymph nodes that contain many lymphocytes. I would walk over to the hospital, which was adjacent to our research labs, don my scrubs and go into the operating room, where the surgeon would drop a tonsil into my cup of saline. Back in the lab, in front of our laminar air flow hood, we would slice up the tonsil with a scalpel and forceps, releasing the individual lymphocytes. From one tonsil we could harvest a billion lymphocytes, which would yield one liter of lymphocyte-conditioned media. Over the course of several months, our cold room gradually filled up with bottles of conditioned media. The next task was to separate the TCGF from other contaminating molecules using biochemical separative methods that Rich Robb, a postdoc had perfected. After we had accumulated enough semipurified TCGF, we injected several mice repetitively, hoping to immunize them so that their spleen cells were capable of making hybridomas that secreted antibodies that blocked TCGF activity. The entire effort was labor-intensive, and it was nerve-racking to take our small quantities of partially purified TCGF to inject the mice. After several failures, Maggie finally identified some hybridomas secreting antibodies that inhibited TCGF activity in the bioassay and blocked radiolabeled TCGF binding, thereby creating the world's first TCGF-reactive monoclonal antibodies. In these efforts it was instrumental that we were skilled in following anti-TCGF activity in the sera of the immunized animals based on both our bioassay and the TCGF receptor assay. This led us to know which of the immunized mice and which hybridomas were producing anti-TCGF.

After we had generated enough of the monoclonal antibodies, we chemically bound them to small beads, which we put in one-milliliter columns. This way we could purify the protein with TCGF activity in a single step. Then we had to prove that we had actually obtained purified TCGF molecules; that is, our preparations were uncontaminated by other molecules. First, we used biochemical assays to search for other molecules and found none. Next we used our precious preparations to determine the amino acid sequence. Proteins are formed by twenty different amino acids bonded together chemically. To determine the sequence of the chain of amino acids, each bond is broken sequentially and the identity of each amino acid is determined chemically. If more than one protein molecule is present, this determination would detect them. To do it we collaborated with the only investigator at Dartmouth who could sequence proteins. We gave him a full milligram of our precious purified preparation. After an agonizingly long month, he reported back that he could not sequence our preparation because it was "blocked," which he explained meant that the first amino acid in the chain could not be determined and thus the next amino acid in the chain could also not be determined and so on. Since technically this was out of my league, I searched around, networked and found Steve Oroszlan at the NCI, who was a seasoned protein sequencer. After several months, Steve found that the first amino acid in the chain was not "blocked," and he successfully determined the first 10 amino acids of the sequence. He also had determined the sequences of several other peptides, short segments of the sequence of the whole protein. I wanted to include this information in our paper to show that our preparation was pure, but I did not want the amino acid sequence to become generally known because it could be used to identify the cDNA, which contained the code for the amino acid sequence of the TCGF protein. Although journal peer reviewers are sworn to secrecy of confidential information in any paper under review, one can never be too careful. For that reason, I withheld the paper at the end of 1982. I did not think that anyone else in the world had the technical expertise to achieve what we had done, so that I felt that we had time. Also, Maggie and I were attempting to clone a cDNA that encoded TCGF in collaboration with Jerry Crabtree. Jerry had been a postdoc with Allan Munck and had moved to the NIH where he was cloning cDNA. By this time, I had a Honda Prelude, which I loaded up and drove back down to

Bethesda. I spent three months with Jerry. By Christmas we had made significant progress and I returned home for the holidays.

We were surprised and dismayed to learn in March 1983 that an investigator from Japan unknown to us, Tadatsugu Taniguchi, had successfully identified a cDNA that encoded a protein with TCGF activity detectable by our bioassay (71). Since we were also trying to identify the TCGF cDNA, this came as a blow for sure. Only several years later did I learn that my very first grad student, Steve Gillis, after leaving the lab sold our TCGF assay cells for $5,000 to a Japanese company, and this company had enlisted Taniguchi's help to clone the TCGF cDNA. Taniguchi had cloned one of the interferon cDNAs in the previous year while a postdoc in Zurich, using the quantitative interferon bioassay to identify a clone that coded for a molecule with interferon activity. He used the same strategy to clone the TCGF cDNA. Without these precious TCGF assay target cells, the Japanese group would never have been successful because the only way to detect TCGF was to use our CTLL bioassay or the TCGF receptor assay to identify which cDNA clone encoded a protein with TCGF activity. Or they could have used TCGF-reactive monoclonal antibodies, but no one else had them.

In science, the only way to maintain one's sanity and to stay in the game is to always look forward. The most important aspect of all these data was that we could now generate and purify large amounts of pure TCGF for the next series of experiments. In this regard we were still way ahead of everyone else. The comparison of our amino acid sequence from our purified protein with the amino acid sequence predicted from the published cDNA was a match. That proved that our protein was not only pure, but this pure protein actually was TCGF, as defined by the bioassay and now by both protein and genetic analyses. This info was crucial. For the first time I knew the true identity of TCGF molecules. We could attribute the TCGF biological activity to a single protein molecule, so I was now comfortable with the new interleukin-2 terminology, and

I used it thereafter. Needless to say, our report describing our successful strategy and tactics — a phrase Max suggested — to generate the first monoclonal antibodies reactive with an interleukin and all cytokine molecules became a classic in the field, cited thereafter by scientists near and far (72).

Soon after we completed these experiments, Ralph Budd, a postdoc who had been a rheumatology fellow at Dartmouth, joined the lab. Like me, he had become disenchanted with medicine. We needed to characterize, and describe the distinctive features of our interleukin-2-reactive monoclonal antibodies because we had discovered four separate antibodies, each of which was different. Ralph took on a project to study the binding characteristics of the antibodies. We knew from our initial experiments that of the four, one worked best to purify interleukin-2, but we didn't know why. To find out, we examined exactly how the antibodies bound to interleukin-2 by performing detailed binding experiments in the same manner that Huey-Mei Wang had with interleukin-2 binding to its receptor. We determined that even though each antibody bound to interleukin-2, it appeared to bind to a different site on the molecule because the antibodies did not compete with one another for binding, and they bound to interleukin-2 at different rates of speed. Also, the bound interleukin-2 would dissociate from each antibody at different rates. All these experiments taught us how to use the antibodies properly. Ralph described his results in a paper that was instructive for many others (73). He also used this information to construct the first immunoassay for an interleukin molecule, an enzyme-linked immunoassay (ELISA), taking advantage of the unique characteristics of how each antibody binds to interleukin-2 (74). Immunoassays differ from bioassays by employing molecules for detection, instead of living cells. As time went on, the immunoassays would take the place of the bioassays for interleukin molecules because the antibodies could be produced and made available commercially.

A portrait of Lynn together with the Olsens,
Roger, chef de cuisine and Meg, dining room manager.

After a few months devoted to getting Café La Fraise up and running, Lynn opened the doors in early 1983. Right from the beginning, it was a resounding success. The menu was classic French, which many people in the Upper Valley had not experienced. Here are some of the offerings from her menu at the time:

◊ *Roast fresh Vermont pheasant with a pear sauce, garnished with a pear beignet*

◊ *Roast loin of lamb wrapped in puff pastry with garlic and sun-dried tomatoes, served with an aromatic rosemary sauce*

◊ *Veal sweetbreads sauteed with mushrooms and brandy and a rich herbed demi-glace*

The restaurant became a favorite of the Dartmouth faculty, especially when they had guests from the big city. They were pleased to show it off as one of the cultural gems of Hanover. I switched taking my visitors from the Hanover Inn to Café La Fraise. Celebrities also found their way to the restaurant. One of Lynn's most memorable guests was Robert Redford, who was in town to receive an award at Dartmouth. Lynn bragged afterwards that she "touched" the famous actor because she shook his hand. Another time Gore Vidal, a leading liberal writer, came for dinner. Dartmouth was quite conservative, so who would also be dining there but William F. Buckley, the leading light of American conservatives? Lynn and her staff worried there could be a confrontation. Not a problem. She placed the ideological adversaries in separate dining rooms, and the evening proceeded smoothly.

Back in the lab, our attention turned to the molecular structure of interleukin-2 and its receptor. With the former, we needed to ascertain its 3-D makeup to gain an understanding of how the interleukin-2 molecule interacted with the receptor. Also, we wondered whether it would be possible to create mutant analogues (variants) that might be either antagonists of interleukin-2 or more potent agonists (stimulants) than interleukin-2 itself. In other words, we wanted to try to improve upon Darwin. Here again Doreen helped immensely. She always got to the lab before me in the morning. When I arrived, she would proceed to ask me about all the things that she thought about over the past 24 hours. This might take an hour or two. One day she told me she had met an interesting investigator from the White River Junction Veterans Administration Hospital in Vermont, just a few miles south of Hanover. Tom Ciardelli was a protein chemist trained to synthesize small proteins from scratch, which requires a lot of sophisticated organic chemistry. He majored in chemistry at Rensselaer Polytechnic Institute in Troy, New York, then got a Ph.D. in organic chemistry at the University of New Hampshire and went on to do a postdoctoral fellowship in peptide synthesis at one of the Max Planck Institutes in Germany. Peptides are short protein chains, made up of fewer than 20 amino acids. At the VA Hospital, he was working on synthesizing thymic hormones — small peptides thought to cause the maturation of T cells in the thymus. This guy was the real deal. I met with Tom and in discussing immunology explained our progress and goals. I said I thought thymic hormones might not be real. Over a short time, it became clear to both of us that if we joined forces, we could help each other immensely. Fortunately, I had enough space in my lab to accommodate him, and he moved from the VA to join us at Dartmouth.

Tom had a network of scientists distinct from my own. One of his friends, Fred Cohen at the University of California at San Francisco, specialized in predicting the secondary and tertiary structures of proteins based upon their primary structure. Here, the critical information is the so-called tertiary structure of a protein because it determines the biological reactivity of the molecule. Protein chemists speak about the primary structure of a protein, which is simply the sequence of amino acids connected by a chemical bond, termed a peptide bond. This sequence then determines the secondary structure, which is the way the amino acids influence their immediate neighboring amino acids. This determines how the protein folds up. Finally, the tertiary structure refers to the final 3-D structure. That 3-D structure is written into the code of the primary and secondary structures, which are determined by the sequence of amino acids. And because there are 20 different amino acids, their sequence is just as varied as any sentence in a paragraph using 26 letters. Predicting how the primary structure of a protein dictates the 3-D structure has been one of the unsolved mysteries of life. Just within the past few years, artificial intelligence (A.I.) techniques termed Deep Mind have made game-changing breakthroughs that promise to predict the 3-D structures of virtually any and all proteins. Tom and Fred teamed up to see if they could put forward a theoretical model of the interleukin-2 molecule. Within a short period of time, they had ascertained that the molecule most likely had four stretches of amino acids that formed alpha helices (the plural of helix, as in the double helix), spiral chains of amino acids. Because this was new information about the structure of an interleukin molecule, our report was accepted by *Science* (75). Tom, being a bona fide chemist and a humble man, joked that *Science* was just a "semi-scientific weekly" and our paper was no big deal. But we knew he was secretly quite proud. In the science hierarchy, the mathematicians are at the tippy-tip top, followed by the physicists, then the chemists, with the biologists, and especially immunologists. At the bottom are MDs, who are not even considered to be scientists, but rather practitioners, and medicine is deemed to be an art form. Tom went on to recruit a team of techs and postdocs who delved into creating mutants of the interleukin-2 molecule, which they studied for the capacity to bind to its receptor and to promote the proliferation of our interleukin-2-dependent CTLL cells, the cytotoxic T lymphocyte lines (76). This became the holy grail of Tom's lab.

One of Tom's recruits to our cause was Howard Holley, a computer specialist who had retired early, around 50 years old, from IBM and was living close by in Vermont. Howard was interested in using computer-generated artificial neural networks, the guts of A.I., to predict protein structure from the primary amino acid sequence. The problem he attacked is fundamental to all living things in that proteins function based on their 3-D structure, which ultimately is derived from the primary amino acid sequence, the string of 20 different amino acids. In turn the sequence of amino acids is determined by the DNA sequence of each gene. That's called the genetic code. Every protein is like a paragraph of text, each of which is different. The 3-D structure of the final protein is determined by how the long protein strand folds up.

At this point, in the mid-'80s, we were all working in a field that was just coming of age. Today neural networks are used extensively in A.I., which among other things, is responsible for voice recognition used in everyday gadgets, devices and applications like Siri and Alexa. Howard was a great colleague for Tom. The computer whiz and crackerjack chemist spoke the same language and learned a lot from each other.

To round out the protein structure expertise, we contacted Ian Wilson, a structural biologist at the Scripps Clinic and Research Foundation in La Jolla. X-ray crystallography was the *sine qua non* (Latin for "cause without which not" and one of my med school mentors Chuck Mengel's favorite expressions) of structural biology. Getting one of structural biology's stars interested in our quest was essential. To explain to him all that we had come to understand about interleukin-2 we invited him to attend one of our meetings at the Minary Conference Center in Holderness, New Hampshire, together with Ellis Reinherz's group We all learned a great deal about how protein structures are determined, which Ellis would put to good use in determining the structure of the T cell antigen receptor. When Ian returned to California, he set to work to obtain crystals of interleukin-2 to use for X-ray crystallography. This method required the use of large (milligram) amounts of pure protein with which to form crystals (table salt and sugar are both common crystals we encounter in everyday life, formed by removing the water from a solution of these substances). Crystals of larger molecules, like interleukin-2, are difficult to generate, and this must be done by trial and error, testing various solutions. If successful, the crystals are then exposed to X-rays so that the atoms on the molecule deflect the X-rays that are detected by spots on X-ray film. It is then possible to calculate their identity and positions on X-ray film. Our contribution consisted of supplying Ian with large quantities of purified interleukin-2. We were moving along swimmingly until 1987 when a report of the "Three-dimensional structure of interleukin-2" appeared in *Science* from a group unknown to us from the University of Colorado (77). Bummer! Such are the vagaries of competitive science. Yet the information that Tom Ciardelli gleaned from this predicted structure greatly aided his protein engineering experiments — or so we thought. Five years later, we learned that this structure was incorrect, thereby negating much of Tom's lab work of those past five years. He'd been relying on the 3-D placement of the sequence of amino acids in order to genetically engineer mutations, then testing the new mutant proteins for interleukin-2 binding and biological activity. Since the 3-D structure was wrong, all the mutants they had painstakingly made were nonsensical. Tom was acutely depressed, to say the least. But this did not deter him. He went back to the drawing board and revamped his plans for the creation of mutants. It was not until almost 20 years later in 2006 that Ian's group came through with the determination of the crystal structure of interleukin-2 bound to the three chains of the interleukin-2 receptor. But that is getting ahead of the story (78).

CHAPTER 8

Structure-Activity Relationships

I WAS INVITED TO A meeting in Osaka in January 1986, my fifth trip to Japan in as many years. By then Lynn had the restaurant running smoothly and I convinced her to accompany me on the trip. We flew nonstop from New York to Tokyo, where we met Ellis to take the bullet train to Osaka. That was fun because Lynn had not spent very much time with Ellis. I was amused watching them interact with one another. They behaved like they were old friends, since both knew me so well. The meeting was small, only two dozen scientists, but it included the stars in the field. I was the only one who brought a spouse, which meant that Lynn stood out at the evening functions, and the geishas in the restaurants were all atwitter about her. I had arranged through my Japanese friends to set her up with tours of some of the finest restaurants in Osaka so she could see firsthand how Japanese restaurant kitchens functioned. She was impressed at how much of the food was prepared ahead of the serving time and the large number of workers in the kitchen for what was truly an assembly-line operation. She also saw they were all male. The only females were janitors who cleaned up at the end of service. Of course, since it was such a male-dominated business, all the restaurateurs were intrigued by their visitor being an attractive woman chef from the U.S.

It was daunting for Lynn to navigate on her own in Osaka with no signs in English. Whenever I toured around the city, I had at least one of the Japanese scientists escorting me. Lynn did just fine all by herself, thank you very much. One day, she went shopping at one of the large department stores and found the top floor had several restaurants, Japanese as well as Chinese and even Western style. After lunch, she needed to use the restroom, so she simply followed the Japanese ladies heading that way as they left the restaurant. She figured her empirical approach to this problem was suitable for the spouse of a scientist.

It was at this meeting when we all heard the news that the Challenger space shuttle exploded just after the launch, a traumatic event. The pictures on TV were sobering, to say the least. So much for science and "modern" technology. We all felt humbled, not quite as smart as before the accident. Also at this meeting, I heard for the first time that Tom Waldmann from the NIH might have evidence for a second chain capable of binding interleukin-2. Takashi Uchiyama had generated the Tac antibody in Tom's lab (79), and we had shown that it could block interleukin-2 binding (57). Tom did not show any data. He just mentioned it in passing during a Q&A after his talk. My ears certainly perked up. Doreen had uncovered some perplexing data on the nature of the interleukin-2 receptor when examining binding data using radiolabeled interleukin-2 or the radiolabeled Tac antibody. She always found 10-20-fold more binding sites on normal activated T cells using the antibody than she did when using radiolabeled interleukin-2. We checked and rechecked our calculations, but such a big difference could not be easily explained by a computational error. Moreover, when exposed to interleukin-2, T cells expressed many more antibody binding sites but only half as many interleukin-2 binding sites. That was strange. Clearly the Tac antibody was recognizing something in addition to the interleukin-2 binding site. Even so, the antibody recognized something to do with the interleukin-2 binding site because it blocked interleukin-2 binding to its site. We went ahead and published these data, suggesting that the interleukin-2 receptor was comprised of more than just the initial chain identified by the antibody (80). But we didn't have a clue what to do next.

Here again, our network came in handy. Frank Ruscetti alerted us to a leukemic T cell line established from a gibbon ape that both produced and responded to interleukin-2 (81). I was immediately interested to find out whether the cell line was growing because of the interleukin-2 it was producing. Could this explain its mechanism of proliferation, and provide the answer to cancer? I performed a series of experiments, taking advantage of the fact that we had already shown that glucocorticoid hormones inhibit normal T cells from producing interleukin-2. I figured glucocorticoids might block this T cell line from producing interleukin-2, which would inhibit the cells from growing. I hypothesized that if interleukin-2 was supplied, the inhibition of proliferation could be circumvented. These new experiments were just like those we had already performed using normal T cells. My hunch proved to be the case in convincing results (82). These data supported the notion of at least one mechanism by which malignant cells could be selected to grow autonomously *in vitro:* they made and responded to their own growth factor. That did not necessarily mean this capacity was responsible for all T cell malignancies, or all cancers, for that matter. Naysayers, with Allan Munck notable among them, pointed out that these findings could be explained by only one cell with this characteristic that had been selected for autonomous growth during the establishment of the cell line *in vitro*. Even so, the reality of these cells growing in response to the interleukin-2 that they themselves produced meant that they had to have interleukin-2 receptors. This also would be the case when we subjected the cells to the interleukin-2 binding assay: it was obviously positive. And yet, when we tested them for Tac antibody binding, they were negative. Definitely strange. Did that mean there was more than one chain capable of binding interleukin-2? The hunt was on for another interleukin-2-binding chain.

I had met Dr. Yukio Sugino, at the 1983 International Congress in Kyoto where he wined and dined me for reasons that were at the time obscure to me. Eventually he told me he had an excellent scientist, Koichi Kato, working under him at the Takeda Pharmaceutical Company who might be interested in joining my lab on a sabbatical year. Takeda is the largest pharmaceutical company in Asia. Dr. Sugino and I discussed whether the best cutting-edge science was being done in academia or industry. Having made the leap from Kyoto University to Takeda and back twice, he argued that the best opportunities were in industry where the resources and, he believed, the people were better. I argued the reverse was true in America. A year went by and arrangements were made for Koichi to take a sabbatical from his duties at the company to spend a year with us in Hanover. I drove to the airport to meet him and his family when they flew in aboard a 20-seater from Boston on a snowy day. They must have been shocked by the sudden difference from their lives in Japan. Happily, they soon settled into a wonderful house in a new development in Quechee, Vermont, just across the Connecticut River from Hanover.

Koichi was another one of those "people persons" I admired. He quickly established a warm relationship with everyone in the lab, especially Barb Atherton, our administrator, who helped him get settled. Barb was a longtime Upper Valley resident and the perfect welcoming agent. Two others also joined our lab at that time. Keisuke Teshigawara, a student of Junji Yodoi at Kyoto University, came on as a postdoc, and Huey-Mei Wang, a Dartmouth Ph.D. student from Taiwan, joined us for her dissertation research. Together, they formed a working group focused on the structure of the interleukin-2 receptor. Koichi as the senior scientist served as the leader of the team; Keisuke and Huey-Mei focused on different aspects of the nature of the receptor. Koichi worked on the protein reactive with the Tac antibody. The problem centered on Doreen's finding of the discrepancy between the number of Tac-binding sites and interleukin-2-binding sites. Koichi's initial experiments uncovered differences between his findings and what had been reported by others. Now, with his considerable biochemical expertise, he performed a series of elegant protein chemistry assays, approaches that could only be interpreted to reveal that the protein existed as both a single membrane molecule, a monomer, as well as a dimer, the term for two molecules linked together (83). However, this information still did not explain the discrepancy between the number of interleukin-2 binding sites detected by the Tac antibody compared with the number detected by interleukin-2.

Keisuke Teshigawara's mentor at Kyoto University, Junji Yodoi, had established another cell line from a leukemic patient, termed the YT cell line, that had a key role in our quest to understand the structure of the interleukin-2 receptor. Junji's father owned a private hospital in Osaka, Yodoi Hospital. The patient from whom this cell line had been derived was being treated at his hospital, hence the name of the cell line, YT. Initially, it was thought to have been derived from an immature T cell because it lacked mature T cell surface molecules like CD3, CD4 or CD8. In retrospect, it was probably derived from a natural killer cell. This became significant because Keisuke performed a series of informative experiments that indicated there were at least two protein chains capable of binding interleukin-2, and when both chains were expressed, a combined receptor was formed that bound interleukin-2 much more efficiently (84). We cloned this cell line and tested the clones to see if we could identify one that bound interleukin-2 but did not bind anti-Tac. Lo and behold, we found such a clone. Then using the interleukin-2 binding assay, Keisuke revealed this new chain bound interleukin-2 with a different affinity (binding strength) than the Tac protein and also different from high-affinity interleukin-2 binding. In addition, only one of these receptor chains could bind the Tac antibody, thereby accounting for the discrepancy in the average number of binding sites on the cell population. At this stage in the saga, it was truly a mystery, and we were determined to get to the bottom of the interleukin-2 receptor structure. Koichi was instrumental in these experiments, sharing his biochemical expertise to help Keisuke and Huey-Mei. Tom Ciardelli was also helpful to Koichi in all matters biochemical. Together they worked out the methods to detect the new chain by first binding radiolabeled interleukin-2 to isolated cell membranes, then monitoring the size of the interleukin-2 receptor chain complex. The new chain had a combined molecular size of 90,000 atoms. Subtracting interleukin-2 itself (15,000 atoms) left a size of 75,000 atoms, significantly larger than the 55,000 atoms that Koichi and others had already found to be the size of the Tac molecule. Thus, we clearly had evidence of an additional interleukin-2 binding chain.

Our stunning finding of another interleukin-2-binding chain, separate and distinct from the Tac protein, was truly exciting. When Koichi and I attended the International Immunology Congress in Toronto in August 1986, I was keen to learn whether anyone else had discovered this new interleukin-2 receptor protein. I was invited to give a plenary session talk on the opening day. I chose to discuss the structure of interleukin-2 (75) that we had discerned with Tom Ciardelli, Fred Cohen and Howard Holley. But I purposely did not talk about the interleukin-2 receptor structural data because at that stage we had not yet submitted our data for publication, and I didn't want to be scooped. The theme of the plenary session was the future of interleukins, so I focused on uncovering the structure activity relationships of interleukins as the first of what I thought would be many hormones of the immune system. I was pleased that no one else reported data on a second interleukin-2 binding chain, and we breathed a sigh of relief. As soon as Koichi and I came home, we all sat down and wrote up both Koichi's and Keisuke's data and sent the papers off.

As the world turns, even though we submitted Keisuke's paper on the discovery of the second chain of the interleukin-2 receptor in August 1986 and it was accepted right away by the *Journal of Experimental Medicine*, it was not published until the January issue in 1987. Meanwhile, a team from the NIH led by Warren Leonard, a former postdoc of Tom Waldmann who had not attended the Toronto meeting, also had data on the second chain, and their paper appeared in *Science* in November, 1986 (85). They had submitted their paper on July 1, 1986, before the Toronto meeting. Moreover, Tom Waldmann's group found similar data when examining the gibbon ape cell line. They submitted their report in September 1986, and it was published that December (86). It certainly took the wind out of our sails since priority is paramount in science; we were third not first with this discovery. On the positive side, the findings of the three papers, published within a month of one another, substantiated the data, thereby establishing unequivocally that there was a second chain of the interleukin-2 receptor. Clearly, our field had become hot indeed.

The other way that we could approach the structure of the receptor was to use the radiolabeled interleukin-2 binding assay to carefully determine how each chain bound interleukin-2. For these studies Huey-Mei Wang proved superb. She had a rocky beginning in the lab when she broke her ankle skating on Occum Pond, next to the med school, just as she joined the lab as a rotating biochemistry graduate student. She was on crutches for three months, so she could not do any experiments, since most of our work was performed standing next to the lab benches, manipulating small test tubes. But this gave her time to read all about immunology in general and interleukin-2 in particular. Huey-Mei's mishap on ice turned out to have an upside for her and us as well because she realized that interleukin-2 and its receptor were at the cutting edge, and she decided to do her thesis work with us. I felt that a perfect project for a biochem student would be to examine how these two chains of the receptor bound interleukin-2. She performed a series of detailed studies examining interleukin-2 binding under steady-state conditions where the target cells expressed only one of the chains vs. both chains. We had already characterized several_different cell lines that expressed the different varieties of the chains capable of binding interleukin-2, so we had unique cells available for study. Huey-Mei became adept at equilibrium interleukin-2 binding assays, where the rate of association equals the rate of dissociation. She found that each chain bound interleukin-2 with a distinct affinity; the new 75,000-atom chain bound interleukin-2 with a tenfold higher affinity than the 55,000-atom chain. However, the icing on the cake was her finding that when both binding chains were expressed on the same cell, the binding affinity was a hundredfold greater, and most important, equal to the interleukin-2 concentrations that promoted T cell proliferation.

For comparison, she also performed experiments to monitor the rates at which interleukin-2 associated and dissociated from the receptor chains. As Allan Munck had taught us, this was a way to check on the results from the steady-state binding experiments. Huey-Mei's experiments were revealing, in that interleukin-2 bound to the 55,000-atom chain rapidly, but also dissociated from this chain rapidly. The reaction was so fast that it was difficult to measure. In marked contrast, interleukin-2 bound to the 75,000-atom chain much more slowly but also dissociated more slowly. The clarifying understanding occurred when both chains were present on the same cell. Here she had found interleukin-2 to bind with the highest affinity. The combined receptor used the fast association rate of the 55,000-atom chain with the slow dissociation rate of the 75,000-atom chain. In other words, the receptor grabbed interleukin-2 quickly and held on. This produced a high-affinity receptor and indicated that the two chains cooperated to form a better binding site than either could alone (87). We speculated at the time that this would become a general way receptors evolved and were selected (naturally, à la Darwin) to be high-affinity binders (88). This was a landmark paper for the immunology community, and even endocrinology. No one had previously encountered such experiments and findings. For sure, immunology had known about experiments like this when immunochemists studied binding of antibodies and antigens in solution, but the data were foreign to most cellular immunologists at the time. Reading the papers now on the molecular characteristics of the interleukin-2 receptor, edited with the assistance of Dr. Landy, I am impressed all over again at our thinking and interpretations. The reports literally sing, and when Huey-Mei graduated, I could not have been prouder of both her and her dissertation.

Eureka!

These findings led to invitations to write review articles for *Science* (69) and the *Annual Review of Cell Biology* (44). To be invited to write a review article for a prestigious journal is a great honor. All our interleukin-2 receptor work was ground-breaking for hormones and their receptors. Exactly how nature had created such high-affinity interactions between molecules had never been discerned. The notion that two receptor chains with lower strengths of binding had evolved to bind the same protein and combined to form a single high-affinity binding site was astounding. I am still not sure how this occurred except by random natural selection. Our data and analyses explained many of the functional observations that had accumulated since the beginning of the 20th century about how molecules interact with one another.

In the middle of all of this, I received word from the NIH that my grant application renewal funding was put on hold, pending the results of a site visit by a team of peer reviewers. This certainly made us pause. I was shocked. We had made so much progress over these past few years. I knew that if we were not re-funded at this critical juncture, it would be a disaster for us. Site visits for large program-project grants that usually involved four or five individual principal investigators were routine, but a site visit for a grant application from a single investigator or a single project was highly unusual. The NIH usually sent about a dozen peer reviewers to visit the lab under question. In retrospect, I can see how the grant review committee had difficulty with my renewal application. I had proposed perhaps a decade of what I thought were the obvious next steps to take, given what we had already uncovered. And these descriptions had to be crammed into the 25-page limit of a routine single-investigator grant. It had to make for turgid reading for the reviewers. By comparison, a program project consisted of four or five individual projects of 25 pages each, a much bigger deal for the reviewers.

We prepared for the site visit with rehearsals for several weeks. Everyone was anxious. Luckily, someone up there was looking out for us. The program person who led the visit team was a seasoned scientist, and the evening before the visit he asked to meet with me over dinner. The Hanover Inn dining room was a staid, old-school kind of a place with well-spaced tables and a hushed environment, an ideal spot for serious discussion. He spent a long time with me. For three hours, we discussed the scientific issues involved in my application renewal. I explained as best I could what I felt were the salient points. At the end of the evening, he gave me sound and pertinent advice on how to carry myself. He stressed that I should not be confrontational, but rather professorial, charming and helpful with any and all questions the reviewers might have. He said he had seen other site visits end in chaos because the principal investigator essentially "lost it" during the visit. Despite a restless night, I was properly prepared for whatever might come.

The session turned out to be rather fun because our team was so proud of the progress we had made, and the experiments planned and detailed in the grant application were so exciting. We were happy to be sharing them with other scientists. Everyone had a role to play in giving a presentation about their projects. In particular, Koichi was a star. It was impressive how clearly he presented his biochemical data on the 55,000-atom Tac protein. Keisuke showed his pathbreaking discovery of the second 75,000-atom interleukin-2-binding chain and Huey-Mei her elegant findings of the consequences of creating a high-affinity interleukin-2 binding site through the cooperation of two distinct low-affinity chains. In a way, these presentations were old hat for us because they were just like the sessions we routinely had at our lab meetings and our gatherings at the Minary Conference Center. We had the site visit in the dark-wood paneled Dean's Boardroom, which was steeped in old-fashioned charm. To top it off, I arranged for the luncheon to be catered by Café La Fraise, which I knew that would be equally impressive. Lynn prepared a buffet of delicious croissant sandwiches.

Long story short, we nailed it.

CHAPTER 9

First Views Into the Black Box

WHILE THE STUDIES WITH HOW interleukin-2 interacts with its receptor were taking place in our lab and others in the 1980s, I began to think about exploring how the signals inside the cell result in the T cell changing its morphology from a small round cell to a large precursor cell, termed a blast, that undergoes DNA replication and mitosis. These were fundamental questions, and they were the ones by which I had started my investigative career. After a decade of research, I figured we were now finally ready to look into the black box. I cannot emphasize enough the importance of choosing your own questions rather than investigating whatever the field deems to be hot at the time. No one else was really focused on the questions I chose, especially in immunology, so the field ahead was untrammeled. Now that investigation of interleukin-2 and its receptors had attracted many competitors, I felt as someone running a small lab that the only way to go was along the road less traveled.

Doreen Cantrell, Martin Gullberg and I were at the Hanover Inn one evening in the early '80s hosting Frank Ruscetti, who was visiting from Bethesda to do some collaborative experiments on the interleukin-2 receptor. Over beers (my favorite was Bass Ale, but I would also go for Samuel Adams), we got onto the subject of the newly discovered genes, termed *oncogenes*, that had been found to cause the transformation of cells in culture to look and act like cancer cells. It seemed to me that these genes, which evidently caused malignant cells to proliferate, might help us explore the interior of T cells directed by the interleukin-2 receptor. Prior to the oncogenes' debut, it's fair to say that cancer was a total enigma. The popular view was that environmental toxins and viruses caused cancer. Oncogenes were the first evidence that cancer might be caused internally by genetic mutations.

To delve into a whole new area, I again went to the literature and took a historical approach to figure out how we might enter this subject. I found the history of oncogenes to be captivating. Oncogenes were first hypothesized to exist in 1964 by Howard Temin from the University of Wisconsin in Madison. His hunch was that they were responsible for cancers in chickens caused by the Rous sarcoma virus (89). This was considered heretical because the virus was an RNA virus. The notion that an RNA virus and not a DNA virus might cause cancer was contrary to the biblical "central dogma" of molecular biology that information in a cell flowed only one way — from DNA to RNA to protein. Temin, the prototypical nerdy scientist, with disheveled hair and large horn-rimmed glasses, was criticized, even ridiculed, at national meetings. But he persisted and modified his theory in 1971 to propose that *cellular genes* normally involved in cellular proliferation could be incorporated by viruses which then could infect other cells and cause cancer. Temin called these normal genes *cellular proto-oncogenes*, which implanted into the scientific psyche for the first time the idea that normal cellular genes could cause cancer (90). It was not until 1980 that Robert Weinberg from MIT and Geoffrey Cooper from Boston University showed simultaneously but independently that DNA extracted from human cancer cells with no evidence of virus involvement could cause normal cells to undergo transformation to look like and behave like cancer cells in culture (91, 92).

This left an overriding question: How do oncogenes cause this mysterious transformation from normal-looking to malignant-looking cells? Unfortunately, the transformation assay Weinberg and Cooper used to identify the genes was another black box.

A cell culture assay that took several days for the cells to undergo a change of their microscopic "looks" and growth characteristics was not amenable to going further as to the molecular mechanisms involved. I felt that perhaps the T cell interleukin-2-promoted proliferation system that Doreen and I had developed (58) might offer a better chance to probe the molecular mechanisms involved, and would at the same time be a way that we could delve into the molecules inside the cells that were activated to be expressed by interleukin-2. It was a perfect time that a new postdoc with just the right expertise entered the lab. Julie Stern had emigrated from Chile to the U.S. in 1973 after the Nixon-Kissinger-C.I.A.-backed assassination of Salvador Allende, the democratically elected socialist president of the South American nation. Julie and her husband, Paul, a pediatrician, were Allende supporters and forced to flee, first to Spain, before coming to America. A decade later, Julie had just gained her doctorate at Dartmouth in a lab focused on the mechanisms of intracellular molecular movement, which made her skilled in molecular biology and molecular genetic techniques. She was about a decade older than most newly minted Ph.D.s. because her education had been interrupted by the right-wing military coup d'etat in Chile as well as taking time off to be a mom.

Since several oncogenes had been identified at that point with the Weinberg-Cooper assay, we decided to use our T cell system that enabled us to stop the cells from proliferating by withholding interleukin-2, then restarting their proliferation by pipetting it back into the media again. In this way, we could synchronize the cells so that most of them would proliferate simultaneously. This also allowed us to separate signals to the cells emanating from the T cell antigen receptor and the interleukin-2 receptor. And it led us to follow the timing of the expression of any oncogenes should they be activated by either of those receptors. As Julie accumulated data, I was excited by the results of her experiments. They showed unequivocally that the receptors for the T cell antigen and the interleukin-2 receptors were distinctly different beasts. Sure, we already knew this because they were capable of inducing different changes in cellular behavior. But in scientific discovery, the more levels of proof, the more acceptance by other investigators. In addition, we had now shown that the timing of the gene expression preceded DNA replication and mitosis. Our data provided confirmation that the T cell-interleukin-2 system was an ideal way to study the molecular mechanisms responsible for normal cellular proliferation. The recognition that a cellular proto-oncogene was expressed (transcribed and translated) by normal T cells synchronously proliferating underscored my original notion that the way to understand cancer was to unravel the molecular mechanisms that control normal cellular growth.

After we had completed these studies and they had been accepted for publication by *Science* (93), I was invited to visit the University of California at San Francisco. I was unaware of their search underway for the head of their microbiology department. I flew to San Francisco and gave my seminar on interleukin-2 and its receptors and our new T cell growth model. During the Q&A afterward, I was challenged by one questioner who said that our findings that the cellular proto-oncogene, *c-myb* (named after the myeloblastosis virus) was involved in cellular proliferation were incorrect. He said that earlier studies had already established that this oncogene was involved in cellular differentiation, which by dogma requires the cells to cease proliferating. Of course, I knew of these earlier studies. It's why we submitted our paper to *Science* in the first place. It was a new finding for this oncogene, and it demonstrated the power of our T cell system to study the role of oncogenes in cellular proliferation. It turned out that my challenger was Mike Bishop, who was one of the major players in the oncogene field in general and the cellular proto-oncogene, *c-myb*, field in particular. He was one of the investigators who had advanced the theory that this oncogene was required for cellular differentiation and proliferation arrest.

After my seminar, I was ushered around to 30-minute talks with faculty members, which were routine for someone like me invited to speak at a university. Most of the people I met were cordial and we had collegial chats about my science and theirs. One fellow, though, was downright confrontational. He was Harold Varmus, one of Bishop's younger colleagues. While clearing off a chair stacked with reprints next to his desk in a crowded and cluttered small office in his lab, he took a shot at me: "So what did you do in this interleukin-2 story? We just had Warner Greene through here talking on the interleukin-2 receptor and he was quite impressive, much more impressive than you."

So it goes in the trenches of scientific research.

I was taken aback by this nasty aggressiveness, but I told him that Greene was a late player to the scene, that I had discovered the interleukin-2 molecule and its receptor and that I had developed a new T cell system with interleukin-2 for the study of normal cell growth regulation. Varmus was not prepared to accept that. He kept badgering me for the entire interview. Needless to say, UCSF did not proffer a job, and for sure, I would not have accepted one even if they had. It was clear to me that neither Bishop nor Varmus wanted me to be a candidate for their department chairman.

The sequel to this episode is that Bishop and Varmus won the Nobel Prize in Physiology or Medicine three years later in 1989 for their 1976 demonstration that a viral oncogene, *v-src*, was present in both virus-transformed sarcoma cells as well as normal cells, thereby proving Temin's cellular proto-oncogene hypothesis (94). For a further postscript, now, more than 30 years later, detailed molecular studies of *c-myb* have elucidated its key role in cellular proliferation. And because it suppresses cellular differentiation, while simultaneously promoting proliferation, it is being considered a target for cancer drug development (95, 96). As I would repeatedly tell my trainees, "Anything is possible in science, but some things take a very long time."

Eureka!

Julie Stern's paper describing *c-myb* involvement in T cell proliferation gave credence to our hypothesis that the T cell-interleukin-2 system was ideal for examining new gene expression induced by the T cell antigen receptor vs. the interleukin-2 receptor. The T cell receptor renders the cells *competent* to receive the *progression* signal from interleukin-2, two terms we borrowed from the cell biologists. These two different receptors regulate coordinated but different aspects of normal T cell growth. Our findings also showed, for the first time, that cytokine receptors on the surface of T cells activate and regulate gene expression of the cells. This demonstration confirmed my notion that the cells of the immune system, like the cells of all the other tissue or organ systems, are carefully regulated by hormone molecules and their receptors. I could now firmly conclude that all multicellular organisms must be regulated in a similar manner and that there are tremendously complex intercellular communication systems comprised of distinct hormones and receptors.

But how did the information generated at the cell surface find its way to the nucleus where the genes reside in the form of DNA molecules? Back then, in the late 1980s, the field that we today recognize as signal transduction (how physical molecular binding on the outside of the cell is converted to biochemical signals on the inside, in the cytoplasm) was in its infancy. Eventually, a complex array of cellular molecules that perform these tasks would be discovered. An inroad into this black box of the interior of the cell was additional viral and cellular proto-oncogenes. They had been detected using the Weinberg-Cooper cell transformation assay and might participate in signal transduction. One such candidate, a gene designated as *raf* (97), was found to encode a cytoplasmic protein that demonstrated enzymatic (catalytic) activity that promoted the addition, or movement, of high-energy phosphate groups onto molecules. These catalytic molecules are called *kinases* (from Greek, *kinein,* to move, the same root used for cytokines, cell movers) and are critical for chemical reactions inside the cell. *V-raf* was first identified in 1983 and *c-raf,* the cellular proto-oncogene, soon thereafter. The designation is an acronym for rapidly accelerated fibrosarcoma inducer. As a result of Julie Stern's experiments, we wondered whether we could use our new T cell-interleukin-2 cell growth model to discover more molecules that are involved in transmitting the growth signal from the receptor at the membrane to the nucleus. From Julie's experiments we knew that *raf* was a likely candidate to explore next.

In order to do so, we understood that we needed biochemical expertise and assays for kinases. Here, my good friend Ellis Reinherz helped once again by putting me in touch with his colleague at Dana-Farber, Tom Roberts, a world-famous authority on the subject of kinases. In this instance, I felt that it would be a perfect project for Antanina Zmuidzinas, a new biochemistry Ph.D. student in the lab. After a preliminary call with Tom, Nina and I drove to Boston to meet and discuss the logistics of the experiments. Tom identified one of his postdocs, Harvey Mamon, to be the primary experimenter from his lab. Nina agreed to handle the cell work and prepare samples we'd send to Boston for their biochemical experiments. Meanwhile, she would retain some of the samples to do her own assays exploring changes in *raf*-encoded mRNA, the messenger molecule that carries the DNA code to the cytoplasm. Nina was perfect for these experiments. She was well-versed in the T cell-interleukin-2 system and skilled in the required molecular biology and biochemistry techniques. To our surprise, she found that stimulation of both the T cell antigen receptor and the interleukin-2 receptor promoted the transcription of the *raf* gene into mRNA. This meant it was different than the regulation of *c-myb* mRNA transcription, which was stimulated solely by the interleukin-2 receptor. However, interleukin-2 stimulation promoted *raf* protein expression and also induced the catalytic kinase activity, whereas the TCR did not promote these fundamental cellular changes. The enhanced kinase activity accumulated over time and culminated just before the synchronized cells replicated their DNA and underwent mitosis. This was exactly what we were looking for. Our report, which appeared in 1991, was the first inkling that cytokines like interleukin-2 promote biochemical changes in the cytoplasm that are associated with signaling normal cellular proliferation (98). Because the expression of v-*raf* was also associated with malignant transformation (cancer), these experiments were some of the first that told us we were on the right track trying to understand the biochemical events regulating normal cellular proliferation in order to uncover what was amiss in a cancer cell.

Thanks to Doreen's experiments and our 1984 *Science* paper, I came to know Arthur Pardee, one of the pioneers of cell growth regulation. Based in Boston at Dana-Farber, he was a legend in scientific circles for his studies in the 1950s on how genes could be turned on and off. He got his Ph.D. at the California Institute of Technology where his mentor was the legendary chemist Linus Pauling. Art was an associate professor at UC Berkeley in 1957 when he took a sabbatical to join Jacques Monod, a master theoretician, at the Institut Pasteur. Being a biochemist, Art refined Monod's experiments that led to the 1959 *PaJaMo* paper, one of the most famous of the 20th century, named after its three authors, Pardee, geneticist Francois Jacob and Monod (94). Their work played a defining role in the discovery of messenger RNA and contributed to the 1965 Nobel Prize in Physiology or Medicine awarded to Jacob, Monod and Andre Lwoff, director of the Institut Pasteur. I always thought that Art deserved a piece of the prize, but the Nobel never goes to more than three people.

Back in Berkeley, Art carried on his experiments with the biochemistry and genetics of bacteria. In 1961 he came east to be the founding chair of Princeton's Biochemical Sciences Department. There, he turned his attention to cancer, especially the molecules responsible for the proliferation of cells. I became aware of his work through another of his renowned papers, this one published in 1974 in *Proceedings of the National Academy of Science* (95). Art established that the growth stimulus, in this case serum, for mouse fibroblasts had to be present for 10-12 hours for the cells to replicate their DNA and divide. Art coined the term "restriction point" (R-point) for when the serum was no longer needed. Some signals had been generated by the serum that seemed to be telling the cells what to do and when to do it. In 1975, he moved to Boston as professor of biological chemistry and molecular pharmacology at the Dana-Farber Cancer Institute and Harvard Medical School.

Because we were both interested in understanding cell growth control, Art found a kindred spirit in our 1984 paper, which placed our breakthrough in the context of 50 years of work on the growth of living cells pioneered by him and many others. He invited me to visit him at his lab in Boston and give a seminar and we hit it off. About 20 years my senior, professorial and soft-spoken, he was a tremendous listener and a probing questioner. He was also a small man, so we had that in common (Being small myself, I like small people.) Another thing we shared was the experience of living in France. From Art's description, his year in Paris was very much like mine two decades later.

Both of us had tackled mysteries unfolding outside of cells. Art had focused on fibroblasts that could be stimulated to proliferate by serum, but he did not know what it was in serum that caused the cell growth. By comparison, we had identified TCGF, which we found was due to interleukin-2 binding to cell surface receptors as the secret agent that directed them to duplicate their DNA and subsequently to divide. For T cells, the focus of my research from virtually day one, my lab pinpointed a single molecule along with its receptor that directed the cells' proliferation. Art and I fit together like the proverbial hand in glove. We found that the concentration of interleukin-2, the density of its receptors and the time that these two molecules interact were crucial in determining cellular proliferation. Next, I wanted to calculate the crucial concentration of whatever molecules might be inside the T cell that triggered replication of its DNA in the nucleus. In order to discuss in depth our common areas of expertise, I invited Art and his group to a meeting with all my lab members at the Minary Conference Center, our favorite Dartmouth retreat in the foothills of the White Mountains. In that peaceful place, we talked for three days about exploring the frontiers of what was happening inside cells.

It was about this time that our lab welcomed another new member. Carol Beadling was destined to play a crucial role in our quest to discover interleukin-2-induced new gene expression. Carol was a graduate of Reed College in Oregon where she had worked on frog interleukin-2 (who knew frogs had interleukin-2?) so that she was attracted to Dartmouth to continue to work on interleukin-2 for her graduate studies. Reed College is a very liberal institution, the polar opposite to Dartmouth College, a bastion of political conservatism. Carol once told me that in Hanover it felt as if she had landed on Mars. But she settled in and began to work on a method that Art Pardee's lab had devised to identify and select for gene expression on a background of so-called housekeeping, or ongoing, everyday, genes. Art had discussed this technique at our Minary meeting, and he helped us get it up and running. This is a prime example of how like-minded scientists assist each other. One hears a lot about competitive scientists keeping things secret from one another because priority is paramount, and second place does not count. Yet scientific friends and colleagues will do anything for one another, and scientists are successful largely because they choose their friends wisely and are loyal, helping them whenever possible.

Art's technique originated with other scientists working with yeast cells. He adapted it for mammalian cells (99). My lab was interested in identifying genes that were expressed as a consequence of interleukin-2 receptor stimulation. We used our T cell system and labeled newly transcribed mRNA molecules with nucleotide building blocks to which a sulfur atom had been added. The sulfur-labeled, interleukin-2-induced mRNA transcripts could then be separated from the unlabeled housekeeping transcripts by means of a mercury column (sulfur and mercury form a strong chemical bond). This resulted in a tenfold enrichment and a 90 percent purification of interleukin-2-induced mRNA, which we used to clone complementary DNA (cDNA) molecules. Carol proved to be adept, learning all of these procedures with help from Kirk Johnson, one of the postdocs in the lab (100). She was tenacious and thorough in her analysis and ultimately identified eight interleukin-2-induced genes, only two of which were previously known and characterized, and only one that was also found to also be expressed after triggering of the T cell receptor. We published these results in 1993 (96). By then the two chains of the interleukin-2 receptor had been found to have sequence similarity with the signaling chains of the erythropoietin receptor, the growth factor for RBCs. (Recall that Torg Fredrickson and I had worked up an assay for erythropoietin while in France 20 years earlier). By the early '90s a superfamily of cytokine receptors had been identified, which included receptors for interleukins 3, 4, 5, 6 and 7 in addition to the receptor for erythropoietin. This was another first. We had devised a gene discovery system that leap-frogged from the cytokine receptors on the outside of the cell directly to the nucleus, thereby bypassing the biochemical signaling pathways in the cytoplasm. This enabled Carol to identify the target genes that governed not only the interleukin-2 effects but also the functional behavior of the many cytokine hormones similar to interleukin-2. In the words of Forrest Gump, it was like opening a box of chocolates. You never know what you're going to get.

CHAPTER 10

interleukin-2 Immunotherapy

ON THE OPENING DAY OF the 1980 Paris International Immunology Congress there was a garden party at the Jardin du Luxembourg. I knew it well because when we lived in Villejuif, we would take the kids there to sail their toy boats in the fountain and ride the ponies around the park, which is in the Latin Quarter on the Left Bank of the Seine. It was a beautiful day. The party was packed with immunologists, picnicking with champagne and petits fours on the lawns surrounding the chateau (where members of the French public were not allowed to walk). I found myself sitting on the grass next to Elizabeth Grimm, a young postdoc at the National Cancer Institute. We exchanged pleasantries, but it wasn't until two years later that I found out what her project entailed.

With all the effort focused on interleukins as newly discovered hormones of the immune system, especially with new data on the structure and function of the interleukin-2 receptor as well as the findings that interleukin-2 induced the expression of specific genes, one would think, or at least hope, that this would lead to new approaches for the immunotherapeutic manipulation of the immune system. Here, the history of immunology becomes pertinent. In the late 19th century, Pasteur predicted that it would become commonplace to prevent most contagious diseases by vaccination. However, with the turn of the 20th century repetitive injections of vaccines or anything foreign came to be deemed dangerous. Immunotherapy was put on the back burner, and immunology was relegated for the first half of the century to the realm of basic scientists, out of the reach of medical doctors. Attempts at immunotherapies took the form of immunosuppression for allergies and autoimmune disorders and, by the 1980s, dampening immunological rejection of organ transplants.

Recall that William Colely introduced the idea of boosting the immune system with injections of bacterial products to treat cancer in the 1890s (5) and in the late 1960s and '70s, Professeur Mathé attempted to refine this approach using the tuberculosis vaccine, bacillus Calmette-Guerin, to nonspecifically augment immunoreactivity to treat leukemia (16). The idea that the immune system could be harnessed to react against cancer was considered foolish by most oncologists, though not Steve Rosenberg, chief of the Surgery Branch of the NCI. Rosenberg had been straining in the traces to begin stimulatory immunotherapy to treat cancer using interleukin-2 as soon as he became aware of our work in the late '70s. Before anyone else, he used lymphocyte-conditioned media in the clinic in the early 1980s. He tried treating cancer patients with such media but had not seen any effects, either beneficial or toxic. Then, in 1985, with the availability of large amounts of genetically engineered interleukin-2 protein manufactured by biotech companies, Rosenberg made the six o'clock TV news and the cover of Time magazine following a press release (which was an unusual way to make scientific information public in 1985) stating that he had treated 25 cancer subjects with interleukin-2 and that half of them had at least a 50 percent decrease in tumor size. That was an unheard-of response. The day of the press release, I started getting calls from media outlets looking for background on this mysterious new molecule, interleukin-2. I told the reporters I thought that the use of interleukin-2 in the clinic was premature.

From his 1985 report, which appeared later in the renowned *New England Journal of Medicine* (97), Rosenberg stated: "The major toxicity due to interleukin-2 administration was weight gain due to fluid retention: 16 of the 25 subjects gained more than 10 percent of their baseline body weight. After several days of interleukin-2 administration, fluid retention often resulted in pulmonary edema, which was associated with "severe respiratory distress ... requiring intubation."

The volunteers in this experiment had to be treated in the intensive care unit, and two of that first group of 25 died of cardiac complications. Those needing a mechanical ventilator were placed into a drug-induced coma and kept breathing mechanically with a tube placed down the throat into the trachea. Rosenberg did not spell out in his report that the fluid retention was secondary to the administration of large amounts of fluid intravenously in his attempts to support falling blood pressures, which progressed as long as the interleukin-2 infusions were continued. An average 10 percent gain in body weight would add 15 pounds to a 150-pound person in only a few days, not a healthy condition.

Although Rosenberg had performed a pharmacological analysis of blood interleukin-2 concentrations after the infusions, he basically ignored his results, and those of our interleukin-2 receptor binding characteristics, both of which indicated that only low interleukin-2 doses would be necessary to activate T cells. Instead, in preliminary dose-finding clinical trials, he used the principles of cytotoxic chemotherapy of dose-escalation until toxicity to establish the highest maximum tolerated interleukin-2 doses for his trial. These chemotherapy principles are based on sound logic: anti-tumor effects are caused by chemicals that directly kill tumor cells. Their efficacies are dose-dependent: the higher the dose the better the kill. However, in this scenario one has to hope that the tumor cells are killed more rapidly than normal cells, which are also killed. That is what accounts for the severe toxicity of chemotherapy.

Since interleukin-2 functions as a hormone, binding to specific receptors on antigen-activated T cells, once the interleukin-2 concentration is high enough to saturate all the receptors, administering even more interleukin-2 should have had no more of an effect, either good or bad. One should be treating the immune system cells, not the cancer cells, which calls for a physiological or natural (hormone-receptor) mechanism. The principles of cytotoxic chemotherapy do not apply. Rather, the principles of endocrinology should have been used as they are when giving insulin to a diabetic. One would not dream of giving too much insulin, correct?

Why did Rosenberg ignore his own pharmacologic data and our interleukin-2 receptor data, forging ahead, despite inducing life-threatening toxicity with his therapy? To gain insight, it is important to understand the differences between surgeons and scientists. Surgeons are "doers" who prefer action to thought; scientists are "thinkers" who use carefully controlled experiments to test hypotheses which they formulate after considerable planning. Rosenberg's training was with Francis Moore, chairman of surgery at the Peter Bent Brigham Hospital in Boston, who favored action over inaction.

But why did these high doses of interleukin-2 cause the blood pressure to drop and fluid to accumulate in the tissues, especially the lungs? The fall in blood pressure accompanied by severe fever, chills, malaise and muscle pain were classic signs and symptoms of septic shock. Answers to these questions would require almost another decade and the discovery of the complete structure and function of interleukin-2 receptors as well as their appearance on cells other than T cells.

In addition to high doses of interleukin-2, Rosenberg also infused leukocytes intravenously, which were taken from each subject and cultured with interleukin-2 for three to four days. He called these cells lymphokine-activated killer (LAK) cells. It turns out that Liz Grimm, a postdoc in Rosenberg's lab with whom I chatted in the Luxembourg gardens, reported that exposing leukocytes from cancer patients to interleukin-2 for two to three days induced these interleukin-2-treated cells to become cytotoxic for their own fresh cancer cells (101). Also, significantly, she wrote that they were T cells. I found this to be strange because I knew that most T cells did not have detectable interleukin-2 receptors, only antigen-activated T cells should have been interleukin-2 responsive. I thought the large majority of T cells should not respond at all.

For his revolutionary therapy, Rosenberg used a combination of lymphokine-activated killer cell and interleukin-2 infusions rather than either of these two treatments alone. He had infused them separately in preliminary studies and had seen no antitumor responses. The tremendous toxicities seemed attributable to the high-dose interleukin-2 infusions, and I had heard him state repeatedly at scientific meetings that "the toxicity *is* the therapy." But the tumor regressions that did occur with the therapies came days, weeks, even months after the infusions had been discontinued. Any beneficial effect seemed to be due to another mechanism, certainly not to direct toxicity to the cancer cells.

Soon after Rosenberg's publicity splash in 1985, a report from Lewis Lanier's lab at the Becton Dickinson Monoclonal Center in Silicon Valley clarified the origin of the cells. Lanier showed that the so-called lymphokine-activated killer cytotoxicity was attributable to natural killer cells, and not to T cells as originally described by Liz Grimm (102). Natural killer cells begot their name after two separate reports from the mid-1970s described the phenomenon of killer cells that did not need to be stimulated by exposure to tumor-specific antigens first, as killer T cells did. They seemed to be able to kill "naturally" (103, 104). In humans, natural killer cells represent a minor subpopulation in the blood, accounting for less than 10 percent of all lymphocytes, while the majority of lymphocytes in the blood are either T cells (75 percent) or B cells (15 percent). Lanier saw all the cytotoxic activity of Rosenberg's cells as a result of natural killer cells, not T cells as Grimm and Rosenberg maintained. Although Lanier's report correctly identified the cell type, it did not explain how or why the high doses of interleukin-2 produced septic shock.

Science is a team sport. Gone are the days of the lone, introverted scientist toiling away in his laboratory. A team leader plays the role of the manager; the hands-on work of the experiments is done by members of the team — usually scientists in training, like junior faculty, postdocs, grad students, and technicians often operating with only bachelor's degrees. After a project is complete, the name of the leader of the group is usually placed last in a report with more than one author when published in journals of science or experimental medicine. This position on the list is termed the senior author. The first author listed is usually the person who performed the yeoman's work on the project and wrote the first draft. Other authors are assigned usually in rank order for work or the importance of their efforts to the outcome of the project. In most instances, a member of the team who performed tasks that did not amount to at least 25 percent of the final product would not warrant co-authorship. In academia, the number of papers authored, and the position of authorship (first or last), is important when the time comes for promotion. This is why it is telling that Steve Rosenberg is listed first on many of his papers, especially if he considered the report to be quite important. This was the case in a series of reports emanating from his group early in the course of his immunotherapy trials. Michael Lotze, the junior scientist who was Rosenberg's right-hand man, actually conducted most of the projects. However, Mike was frequently listed as second author and not last (senior) author either.

This paradigm held true in a follow-up paper, again in the *New England Journal of Medicine* in 1987 (105). The Rosenberg team still did not report whether interleukin-2 alone might be just as good as the combination of lymphokine-activated killer cells plus the high-dose interleukin-2. They reported their results with 106 subjects who received the combination and 45 subjects who received only interleukin-2. An objective response to the combination was reduced in half compared with the original report (to about 25 percent). The results with high-dose interleukin-2 alone were similar. And almost all the patients experienced severe toxicity. For me the question remained: if the toxicity was the therapy, why did all those who experienced the toxicities not also experience cancer disappearance?

From the small numbers of individuals treated, it was still not clear whether the combo might be slightly more efficacious than simply high doses of interleukin-2. Two years later, the Rosenberg team reported their collective experience with 652 subjects (106), and after that the results of a randomized controlled trial in 1993 (107). All these data yielded essentially the same conclusions; there were no significant differences between the two forms of therapy, both the efficacy as well as the toxicity, and with more subjects, the complete response rates were halved again, this time to only 5 to 10 percent. After almost a decade, the lymphokine-activated killer cell infusions were finally discontinued. I believe it is significant that had Rosenberg been based at a university rather than the NCI, he would never have been permitted to carry out these trials over 10 years because university Institutional Review Boards would have never approved subjecting people to such toxic treatment for such a small percentage benefiting.

The toxicity of Rosenberg's high-dose interleukin-2 therapy was finally explained after two decades of scientific detective work with inflammatory cytokines, initiated by two groups of investigators, one from Yale led by Byron Waksman and performed by Ph.D. student Nancy Ruddle, and another group led by Gale Granger from UC Riverside. They first reported an activity termed lymphotoxin in 1968 (108, 109). Then Lloyd Old's group at Sloan Kettering discovered an activity they termed tumor necrosis factor, first reported in 1975 (110). Both activities were shown to kill tumor cells *in vitro*. The molecules responsible for these activities were subsequently identified by detailed biochemical techniques in the mid-1980s. The cDNAs encoding them were then identified and cloned, thereby allowing for the production by genetic engineering of large enough quantities of pure protein for clinical trials by the late '80s.

The race was on. As soon as pure tumor necrosis factor molecules became available for clinical use, investigators from major cancer centers worldwide rushed to be among the first to use them. Tumor necrosis factor got its name not only because it could kill tumor cells *in vitro*, but also because it caused the necrosis (death) of large, visible solid tumors *in vivo* within 24 hours when injected into mice-bearing tumors (110). Reports ascribed both the capacity to cause cell death *in vitro* and *in vivo* to its direct tumor cytotoxicity. However, it's more likely that the signs of vascular-circulatory collapse in septic shock that occurred right after contact with tumor necrosis factor could have compromised the blood supply of highly vascular tumors, and a falling blood pressure could cause fatal oxygen deprivation of the tumor cells.

I met Jordan Gutterman when I was working with Georges Mathé in Villejuif in 1973-4. Mathé was at the height of his celebrity. It seemed like every up-and-coming cancer immunotherapist passed through his institute. Gutterman arrived from the MD Anderson Hospital to give a talk at one of our *grand dejeuners.* After the lively luncheon, we retired to Mathé's office *a prendre un café*, which of course was an espresso. The office was made for this kind of informal chat, in that along with Mathé's large desk, there was an area with lounge chairs and a coffee table. He was gracious and seemed interested in everything Gutterman had to say about what was going on in Houston, as well as other centers regarding new approaches for cancer immunotherapies. After the visitor from Texas left the room, Mathé turned to me and said, *"Je ne crois pas un mot ce qu'il dit"* (I don't believe a word he says). As a young stagiaire just beginning to learn the ways of the world I was quite shocked, first that he would think it and second that he would say it.

Gutterman's group at MD Anderson was one of the first to report administering purified tumor necrosis factor in a phase one, dose-finding, toxicity study of cancer subjects (111). For phase one trials, the idea is to begin with very low doses with a small number of subjects, usually only four, and then if there are no toxicities, to increase the dose with successive subjects until side effects occur. They found a single intravenous injection to be well tolerated and no evidence of circulatory collapse. A group from Roswell Park Memorial Institute in Buffalo soon reported a similar dose-escalation study (112). They administered a one-hour intravenous infusion and found that high doses led to a fall in blood pressure. Subsequently, at Dana-Farber, a 24-hour continuous intravenous infusion (113) was found to cause a fall in blood pressure and unacceptable toxicity.

These first human trials aimed to see if tumor necrosis factor would kill tumor cells directly as its primary role. However, Bruce Beutler and Tony Cerami from the Rockefeller University had determined that tumor necrosis factor interacted with receptors on the surface of many body tissues (114). This threw a monkey wrench into the idea that the substance was a potential anticancer agent. With the new *in vivo* data (115) that acute high doses of tumor necrosis factor could lead to falling blood pressure while chronic lower doses caused cachexia (generalized wasting), a new notion took shape: that this substance given to people with cancer was not simply an anticancer agent but a newly discovered physiological hormone that regulated normal bodily functions via binding to receptors and activating many cells in the body (115). Because of the toxicities of its administration, it appeared that tumor necrosis factor and lymphotoxin were two of the primary mediators of inflammation, both acute and chronic. That dimmed enthusiasm for their potential to be anticancer agents. Subsequent studies showed that one of the primary effects of tumor necrosis factor and lymphotoxin was on capillaries, the smallest blood vessels. It made them leak plasma into the tissues. On a local scale, this effect is beneficial; it allows active ingredients of plasma like antibodies and leukocytes access into the tissues when there is an infection or tissue injury. The effects of lymphotoxin and tumor necrosis factor cause the signs and symptoms that are recognized as inflammation (red, hot, swollen and sore). On a systemic level, large leakage of plasma drops blood pressure to lethal levels, and that is recognized as sepsis or septic shock.

These considerations were underscored by another study by Tony Cerami, this time with Charles Dinarello of interleukin-1 fame in collaboration with clinical investigators from Brigham and Women's Hospital (116). They administered bacterial lipopolysaccharide (LPS), which was known to be the prototypic agent to cause septic shock, to normal, healthy volunteers in a well-controlled research center-hospital setting. They found that the timing of the onset of all the cardinal signs and symptoms of bacterial LPS-induced septic shock *followed* the appearance of plasma tumor necrosis factor. The symptoms did not precede detectable tumor necrosis factor. This was the Beutler and Cerami mouse LPS experiments reproduced in humans. It identified for the first time a physiological molecular mechanism deemed responsible for the classic clinical characteristics of septic shock. The mechanism employed cytokines produced by the body in response to a bacterial product. In other words, bacterial LPS is *not* a toxin produced by bacteria (an endotoxin) as was originally thought by German bacteriologist Richard Pfeiffer in 1894. Rather, bacterial LPS causes the release of tumor necrosis factor and lymphotoxin, which are the endotoxins. It took almost a century to discover the actual causes of inflammation. Fast forward to 2022 and the Covid pandemic, the symptoms ascribed to the infection are not due to the coronavirus itself, but to the host inflammatory response to the virus, which consists of the pro-inflammatory cytokines. And leakage of plasma from the lung capillaries into the tiny air spaces leads to hypoxia and, if not stopped, death.

The toxicities resulting from Rosenberg's high-dose interleukin-2 therapy were remarkably reminiscent of LPS-induced septic shock. As Rosenberg pushed on, I always felt it was crazy to use such high interleukin-2 doses; there just had to be a better way. And I did not understand why high doses were so toxic. I believed we needed to find out more about interleukin-2 receptor structure and expression on different cells, as well as how the interleukin-2 receptor signaled the cells. Even though I personally had not been involved in Rosenberg's experiments, the background of these events influenced my thinking and eventually caused me to shift my attention from the lab to the clinic.

CHAPTER 11

The Third Interleukin-2 Receptor Chain

As the 1980s progressed, Lynn was preoccupied with her restaurant and I with my science. Meanwhile, Kerry was becoming a star goalie on the Hanover High varsity hockey team. When I told him that he should think where he should apply to college, he said, "Dad, I don't have to worry about college, I'm going to play for the Bruins." That would be Boston's professional hockey team. I suppose this should not have been a shocker for me since whenever I inquired about his homework, he always said that he had done it already in study hall. I had to point out to him he had inherited my size. "Have you seen those guys? You're only five-seven. They're a lot taller and weigh 100 pounds more than you." I told him, "When I was your age and had some experience with summer jobs, I realized that it was going to be much better for me, being so small, to use my mind rather than my body." I'm afraid I was not convincing. One of my regrets growing up was that I did not listen to my father, especially when he advised me to take Latin and Greek. I was going to be a doctor, so what did I need dead languages for? Of course, in medicine and science, almost all the terms have Greek or Latin roots.

Kerry graduated from Hanover High School in 1987. After the commencement ceremony as we were posing for photos with his classmates, Kerry turned to Lynn and me and said, "You were right." To which we said, "About what?" Whereupon he said, "I should have studied." He had graduated with a B-minus average and saw many of his classmates moving on to prestigious schools. He now realized that he needed to play catch-up. So instead of going directly to college, he applied to prep schools for a postgrad year. His athletic prowess in soccer and hockey made him attractive to Phillips Academy in Andover, Massachusetts, just outside of Boston. Lynn and I felt he still had a lot maturing to do, so it seemed like a good fit.

Over the summer, while rigorously training, he fractured one of the small bones in his foot. The X-ray showed a split right down the middle of the bone. This is a common fracture for both basketball and soccer players, which are obviously two sports that put a lot of pressure from the leg bones onto the small foot bones. So instead of Kerry padding the Andover soccer and hockey teams, he would spend eight weeks on crutches with a non-weight-bearing cast followed by a few more months with a single walking crutch. He had signed up for some difficult courses, including calculus, modern English literature and linguistics, and for the first time he studied. Lo and behold, he achieved outstanding grades for his first term, straight As. I figured finally that there might be hope for this lad.

I thought it would be good for Kerry to gain some experience out in the world. After all, Hanover was an extraordinary, picture-perfect place, and neither one of our boys had ventured very far afield. My many trips to Japan had given me a deep appreciation and admiration of that country. I figured that Kerry's generation was going to be influenced by Japan even more than mine. I thought to myself that maybe we could find a lab there for him to spend some time as a college student. With that in mind, I planned a two-week trip for both of us to Japan in the spring after his year at Andover. Along with being tourists in the major cities and main islands, I would give talks about the state of the art regarding interleukin-2 and its receptor at several institutes and research companies. I told my contacts that we were looking for a lab where my son might work for a year while he was in college. Our trip would have been perfect, except that just before we were to leave, Kerry suffered another stress fracture in his other foot. This put him back on crutches. He wasn't going to let that stop the trip, so off we went. We flew from New York to Tokyo, then took the bullet train to Osaka, where we spent time with Koichi and his family. Koichi took us to Shikoku, the smallest of the four main islands, a picturesque rural community, largely untouched by modern development. It was as if life there had not changed since ancient times. Next it was off to Kumamoto University, on Kyushu, Japan's southernmost island, where we visited Kyoshi Takatsu of interleukin-5 fame. Kyoshi was another member of the Japanese contingent that focused on the new field of interleukins. Interleukin-5 was one of the interleukins produced by activated T cells that in mice was responsible for helping B cells produce antibodies. He took us to visit one of the active volcanoes on the island. I still have a sulfur rock I picked up there.

In Kyoto we visited Junji Yodoi and Takashi Uchiyama who hosted a dinner after my talk at Kyoto University and Kerry had a chance to talk with the students. Kyoto is the city that I visited on my first trip to Japan. Untouched by World War II, it was once the capital of Japan, so it has the look and feel of old Japan. The Zen gardens at the many Buddhist temples are especially impressive. A special treat for me in Kyoto was to meet Shinpei Kasakura, who in 1965 first reported the *blastogenic factor*, which was the first description of a factor in lymphocyte-conditioned media that promoted lymphocyte proliferation, and thus, the forerunner of interleukin-2 (33). Shinpei continued his research at Kyoto University after his postdoc at McGill University in Montreal. By this time, almost 25 years later, he had a thriving hematology practice. He was pleased that we had grasped the baton and brought mitogenic factors into prominence.

Back at Osaka University, we visited Tadatsugu Taniguchi, who had cloned the interleukin-2 cDNA, and Tadamitsu Kishimoto, in whose lab Toshio Hirano first purified and characterized the interleukin-6 protein and cloned the interleukin-6 cDNA, another of the molecules responsible for stimulating B cells to produce antibodies and one of the major interleukins responsible for inflammation. I gave my lecture to their combined groups. The room, which accommodated about 100 people was packed. During my presentation, I learned that Taniguchi had just cloned the cDNA encoding the interleukin-2 receptor beta chain but had not published it yet. I could tell by the reactions from the audience to some of my statements. Since we were also in that race, that was certainly not good news. I'll never forget Kerry's face at the back of the room when we heard the news.

In Tokyo, we visited Tomio Tada, one of the first young postdocs who came to the U.S. to train with Kimi and Terri Ishizaka at Good Samaritan Hospital in Baltimore. The Ishizakas had achieved international recognition in 1967 for discovering immunoglobulin-E, the antibodies responsible for allergic reactions, and they served as trainers for a succession of Japanese immunologists in addition to Tomio Tada, including Kyoshi Takatsu, Junji Yodoi and Tadamitsu Kishimoto, to name just a few. These are the Japanese immunologists who really put Japan on the global immunological map during the '70s and '80s. I had met the Ishizakas after a lecture I gave at their labs just before our trip to Japan, so that I felt I now knew the core of the team that helped make Japan well known in immunological circles.

The highlight of the trip was our visit to Kazuo Sugamura at Tohoku University in the city of Sendai, northeast of Tokyo (Tohoku means northeast) on Honshu, the main island. Two decades later, Sendai would be the site of the 2011 Tohoku earthquake and tsunami. With the surrounding hills reminiscent of the Appalachians of New Hampshire, Kerry and I felt right at home. Kazuo took us up into the mountains to a traditional Japanese inn, called a ryokan. We each were given a kimono to wear. The rooms were spacious with tatami mat floors and sliding doors but no furniture. After a bath in the hot-spring pool, we were treated to an elaborate multicourse Japanese dinner, served by geishas who entertained us with traditional music and singing. By this time, both Kerry and I were proficient with chopsticks, at least for westerners. Accompanying us was a molecular biologist from the group, Masa Nakamura. Little did we know it at the time, but both Kazuo and Masa were checking out Kerry as much as Kerry was getting to know them. The upshot of this trip was that Kerry eventually became a biochem major at Connecticut College and enrolled in a Japanese course. He also joined the crew team, which gave him an outlet for his athleticism and unbridled energy. Our Japan trip was successful in that a few years later, Kerry took a year off from his studies at Conn College and joined Kazuo and Masa in Sendai.

Breakfast at a ryokan in Sendai

Kelly presented other challenges. When he was in the eighth grade, his teacher called and requested a meeting with us. We had no idea what this might be about, but it sounded ominous. A few days later, his teacher sat us down and told us: "I have lost control of my class to your son." She said she could not do anything unless Kelly approved of it, no matter what it was, and the whole class looked to him for guidance. Kelly had trouble learning to read in the first and second grades. We held him back a year, but that had no effect. We finally had him tested by a learning expert who told us that he had classic signs of dyslexia. Dyslexia is a learning disability that becomes manifest usually when a child first starts to learn to read in first and second grade. The key word here is disability, and like many physical disabilities, it is lifelong. It is often found to be inherited. Lynn was listening in while Kelly was being tested, and she realized that she had very similar symptoms. It was an eye opener for her. In all her schooling, she never understood why reading was such a problem and no one said anything about dyslexia. Of course, learning disabilities had not yet been recognized. She compensated by developing her interpersonal skills, and she thrived in all things social. The hospitality business of restaurant work was a perfect way for her to use her skills where she could shine.

For Lynn, Kelly's difficulties learning turned on the proverbial light bulb. She vowed then to help him so that hopefully he would be given a chance to learn how to learn. We were fortunate to live in a college town where the focus of the community was education. We found a super special education teacher for Kelly. Working one on one with her he made progress in primary school. He gradually learned how to get around his disability, but learning by reading was still a struggle. He could not keep up with his classmates and he was bored. One of the other hallmarks of dyslexia is a normal intelligence. Even though they might appear dull on first glance, people with dyslexia can usually find ways to circumvent their boredom. One of the most common ways is to become the class clown, and this is how Kelly was able to take over control of his class. He was popular among his classmates because he was entertaining and irreverent of authority, a personality trait that Lynn was sure he received from me. She was probably correct. I think I obtained that trait directly from my mother, who was always skeptical of authority figures dictating how she should conduct her life.

It was obvious that we needed to change Kelly's modus operandi. Networking with other people with dyslexia and the professionals, we located a superb private school in Litchfield, Connecticut. At the Forman School, all the students and many of the teachers were dyslexic. It was a tearful separation when we drove down there in the fall of 1985 for Kelly to begin ninth grade. He later said that it was the best thing that we ever did for him. He thrived and felt that the first time in his life, everyone was just like him. At Forman, in addition to many new friends, Kelly delved into art, discovering he had real talent. As time went on, this aspect of Kelly's makeup became life-defining. My mother took up art later in life after having discovered that she also had talent when she enrolled in a painting class at her church. Between the ages of 60 and 80, she produced more than 400 paintings. We have several of them, including a seascape from a postcard that we sent to her from Brittany just before we returned from France. Now I find myself following her example. After several years of classes and working with different media, I've settled on pastels and focused on portraits because I always felt that they were the most difficult. I've found that millimeters make a huge difference in whether the likeness is true or not. Behavior is one aspect of our lives that has been debated over the years by psychologists, centering on the question of nature vs. nurture: Which is of greater influence on an individual's behavior, inherited traits or those that can be traced or taught through experience? As time goes on and new genetic techniques are applied to psychology, there's more and more evidence that favors nature over nurture in personality development. In large part, the nature in what is described as human nature accounts for the importance of the family unit and our tribal and societal groups. It's in the genes.

Michael Palladino was a young Ph.D. student, first trained at NYU by Jeanette Thorbecke, an established senior immunologist. From there, he went on to work as a postdoc with Lloyd Old at Sloan Kettering in the early 1980s where he learned all about tumor necrosis factor. Then he landed a job with Genentech in San Francisco, one of the first biotech companies founded in the '70s, even before the passage of the Bayh-Dole Act in 1980 that established the rules of patent ownership of biological discoveries funded by the federal government. In the Department of Molecular Biology and Immunology, Palladino lobbied hard for the young biotech firm to apply its biochemical and molecular biology expertise to cytokines, particularly emphasizing tumor necrosis factor and lymphotoxin as biological molecules with therapeutic promise. Intrigued by the similarities of the toxicities of high-dose interleukin-2 therapy and tumor necrosis factor administration, Palladino collaborated with a clinical group at the UC San Francisco that was conducting interleukin-2 studies using Rosenberg's high-dose protocol. They found that interleukin-2 given as a high-dose intravenous injection every eight hours led to the appearance of readily detectable plasma tumor necrosis factor, lymphotoxin and interferon in the blood (117). Similar findings were reported by James Mier and Charles Dinarello at Tufts-New England Medical Center (118). Systemic toxicity due to high doses of interleukin-2 correlated with interleukin-2-induced production of these pro-inflammatory activities. However, the cellular sources of all these cytokines remained obscure. Where were they coming from?

Around this time, I received a call from Jerry Ritz from Dana-Farber, one of Ellis's best friends. Jerry was treating leukemia patients using bone marrow transplants just as Mathé had started to do in the late 1950s. Jerry told me that he wanted to try interleukin-2 therapy to help boost the immune systems of his transplant patients because he was losing some of them due to infections in the immediate post-transplant period when they essentially had no immune systems. I said, "I don't know how exactly to accomplish that, but I do know what not to do, and that is to give the huge interleukin-2 doses like Rosenberg is using." Based on our 1984 *Science* paper (58), I knew the important determinants of interleukin-2 effects on T cell proliferation were threefold: the interleukin-2 concentration, the interleukin-2 receptor density and the duration of the interleukin-2 receptor interaction. Thus, I felt we should give low interleukin-2 doses that would saturate the interleukin-2 receptors and administer it longer than for just a few days as Rosenberg had done. Also, since we had just discovered the new interleukin-2 receptor beta chain, I felt we should examine normal leukocytes again to see which cells expressed the new receptor.

We embarked on a series of experiments performed by one of Jerry's postdocs, Michael Caligiuri, and Nina Zmuidzinas from my lab (119). We examined the expression of both chains of the interleukin-2 receptor proteins as well as the expression of their mRNAs. To our surprise, we found that only about 2 percent of natural killer cells expressed the interleukin-2 receptor alpha chain, but as many as 50 percent expressed the beta chain. And since the beta chain had a hundredfold lower affinity for interleukin-2, it became obvious, at least to me, why Rosenberg's high-dose interleukin-2 administration might cause the severe toxicity. Low interleukin-2 doses that would result in low interleukin-2 concentrations throughout the body would be expected to selectively stimulate only the small fraction of natural killer cells (about 2 percent) that expressed both interleukin-2 receptor chains. And I felt that those low interleukin-2 concentrations should not cause toxicity because not many cells expressed high-affinity interleukin-2 receptors.

That is what we found in our phase one, safety clinical studies of the administration of low-dose interleukin-2, which we administered as a continuous intravenous infusion over three months in an outpatient setting (120). This landmark 1991 study was spearheaded by Mike Caligiuri, and the aim was to find the maximum nontoxic daily interleukin-2 dose. To accomplish this, a catheter was placed into an arm vein and the volunteers were outfitted with a pump that would deliver the prescribed dose on a continuous, daily basis for three months. Beginning at a dose of five micrograms, we eventually found the maximum nontoxic daily dose to be 60 micrograms. If we exceeded that dose, fever and generalized malaise occurred. By comparison, Rosenberg used 4,200 micrograms a day (70 times greater), which he gave as intermittent IV push doses three times a day. And by comparison to the severe high-dose toxicity, our daily continuous low-dose intravenous infusions produced no serious, grade 3-4 toxicities. To our delight, there were measurable immunological effects, the most dramatic being a marked increase in the concentration of circulating natural killer cells over the three-month treatment interval. This amounted to an increase in the percentage of natural killer cells among all lymphocytes in the blood from an initial 5-10 percent to as much as 75 percent. In addition, there was a progressive increase in the efficiency of natural killing, Rosenberg's so-called lymphokine-activated killer cell phenomenon. We had discovered a way to achieve immuno-stimulation without Rosenberg's severe toxicity, which he had concluded *is* the therapy.

Just one year later, in 1992, it was a huge surprise when Kazuo Sugamura from Sendai published two papers, one on the discovery of a new third interleukin-2 receptor protein (121) and another on the discovery of a cDNA encoding this chain (122), which was then termed the gamma chain. I thought it was great that they discovered the third chain. I was especially happy that I had chosen not to enter the race for it, particularly because Kerry had taken a year off from college and was spending it with Sugamura's lab, with Masa Nakamura serving as his primary mentor. It turned out that Masa was the coach of a crew team at Tohoku University, and they drafted Kerry to join them. He had a Belgian girlfriend and taught English in an evening class to young Japanese students. He was having a wonderful year.

It took a decade, from 1982 to 1992, to reveal that the interleukin-2 receptor is comprised of three distinct protein chains. Of considerable interest to us, by itself the third chain was found to bind interleukin-2, but with a very low affinity, the lowest of the three chains. However, the gamma chain combined with the beta chain created the very slow interleukin-2 dissociation rate, producing the high affinity three-chain receptor (123). Soon, the gamma chain was found to be involved in receptor signaling of several other interleukins, including interleukin-4, 7, 9, 13, 15 and 21 (124). As new interleukins were discovered on the basis of their molecular characteristics, they were numbered according to their order of identification. Moreover, one of the most devastating congenital immunodeficiencies, known as X-linked severe combined immunodeficiency (SCID, termed "combined" because T cells and B cells were both missing), was found to be a result of loss of function mutations in the gene encoding the gamma chain, thereby solidifying the importance of interleukins, and especially the gamma chain in the normal development and function of both T cells and B cells. These findings opened the way to identifying other congenital immunodeficiencies due to mutations in the genes encoding the interleukins, their receptors and intracellular biochemical signaling molecules. What are now called "inborn errors of the immune system" have become one of the rapidly expanding areas of immunology — all attributable to the X-SCID-gamma chain mutation as a model.

Mike Caligiuri moved to Roswell Park Memorial Institute in Buffalo to begin leading his own group in 1993. There he extended the studies on the immunological effects of low-dose, continuous intravenous infusion interleukin-2 therapy and showed that low doses resulted in serum interleukin-2 concentrations that could saturate high-affinity interleukin-2 receptors comprised of all three receptor chains and that promoted the expansion of circulating T cells (125). The discussion in our report concluded: "The design of our interleukin-2 regimen was based on the pharmacokinetics of interleukin-2, the binding affinities of the interleukin-2 receptor and the known distribution of the interleukin-2 receptor on lymphocyte subsets. Further rational clinical trials with interleukin-2 and other cytokines could be undertaken using these data and the principles of endocrinology and immunology rather than cytotoxic chemotherapy."

Soon thereafter in 1994, a new cytokine, interleukin-15, was discovered (126). It was very similar to interleukin-2 as it was shown to bind to interleukin-2 receptor beta and gamma chains after binding to its own distinct alpha chain. In contrast to interleukin-2, which is produced solely by antigen-activated T cells, interleukin-15 was found to be widely produced by a variety of tissues, including placenta, skeletal muscle, kidney, bone marrow cells and, notably, LPS-activated macrophages. Of particular interest with regard to the toxicity of high-dose interleukin-2 therapy, Mike found that the interleukin-2 receptor beta and gamma chains expressed by the majority of natural killer cells transmitted signals from interleukin-15 as well as interleukin-2, enhancing natural killer cell production of interferon gamma (127). However, without the interleukin-2 alpha chain, much higher interleukin-2 concentrations (about a hundredfold higher) were necessary to stimulate the majority of natural killer cells. These results were especially important because in contrast to interleukin-2, interleukin-15 was found to be produced by LPS-activated macrophages, and natural killer cell-derived interferon gamma was already known to be a potent activator of macrophages, enhancing their production of all of the pro-inflammatory cytokines such as tumor necrosis factor, lymphotoxin and interleukin-1 (128).

There goes the proverbial lightbulb again. I finally understood why Rosenberg's ultra-high-dose therapy was so toxic. By using such high doses, interleukin-2 therapy sets in motion a circular systemic inflammatory response promoted by natural killer cells via the high concentrations of interleukin-2, which stimulate the interleukin-15 receptors on natural killer cells. The discovery that interleukin-15 also uses the beta and gamma chains of the interleukin-2 receptor provided the missing explanation for the high-dose interleukin-2 toxicity. Since there are one billion natural killer cells circulating in the blood, the high concentrations of interleukin-2 stimulated a sudden release of interferon gamma. And interferon gamma is known to activate macrophages, thereby providing for more than enough production of systemic inflammatory cytokines to lead to high fevers, vascular collapse and septic shock. Most of the published reports on this subject, which are voluminous, agree that the macrophage-derived pro-inflammatory cytokines and the natural killer-derived cytokines (especially interferon gamma), are responsible for the vascular capillary damage and subsequent vascular leakiness. Thus, just as with the septic shock caused directly by bacterial LPS stimulation of macrophages, the major sources of the pro-inflammatory cytokines are the macrophages, while in the case of high-dose interleukin-2 therapy, the initiator is the high-dose interleukin-2-activated natural killer cell that produces interferon gamma, which in turn promotes pro-inflammatory cytokine production by macrophages. The result is the same no matter the initial cellular source, and the pro-inflammatory cytokines are also the same, in particular, tumor necrosis factor, lymphotoxin, interleukin-1 and interleukin-6.

By 1995 after a decade of Rosenberg's high-dose interleukin-2 therapy that caused 100 percent of the patients to undergo severe toxicity, only five percent received a benefit. Of course, for the five percent this outcome was great, but for the remaining 95 percent there was no benefit, only severe toxicity. I always felt that Rosenberg did more harm than good to the cytokine field, and especially to interleukin-2, by forging ahead before anyone really understood the complexity of the immune system. The question remaining unanswered was the difference between those whose cancers responded vs. those whose cancers did not respond. After all these data, one could not conclude as did Rosenberg that "the toxicity *is* the therapy." The bottom line is that the immune system is regulated by a complex network of hormonal cytokine molecules that direct cellular behavior by interacting with cell surface receptors, which in turn regulate the expression of genes controlling additional cells. Accordingly, trying to intervene therapeutically, for example with high-dose interleukin-2, in such a complex system is fraught with consequences that were not and never could have been anticipated.

The hopeful news after such dismal results was that I now understood that one could give continuous interleukin-2 therapy to augment the immune system without toxicity.

CHAPTER 12

From the Top of the World to the Capital of the World

WHILE OUR STUDIES ON THE molecular genetics of cellular proliferation and differentiation were ongoing, it became clear to me that the experimental molecular techniques were becoming more and more labor-intensive. I needed more labor. I knew I would need to beef up the team. To keep the grant writing to a minimum, I had purposely kept my team small, with only a half dozen people at one time. We were able to stay ahead of our competition by being fast (and smarter). Of course, it helped that we had created our own crucial cells and purified molecules, which for several years no one else had. I realized I had to change my modus operandi and build a bigger group if I wanted to follow my dream of cracking open the black box of cancer. But the pace of discovery would be slower. Experimental molecular work took weeks and months to come to fruition, instead of hours and days, as it did when it was primarily cell culture work. Often, the experimental results were qualitative rather than quantitative, and to be sure they were correct, most experiments had to be repeated and repeated. I began to think about additional avenues that we might pursue in order to increase our funding and build the lab to attack these new vistas. I also had a new goal germinating in my mind: to bring new therapies to immunological diseases made possible because of our understanding the hormonal immune system at the molecular level.

My small study in our house in Hanover was paneled in cedar with a fireplace that I continuously stoked when the cold weather descended upon us. It seemed like I spent more time stoking the fire than I did writing. My desk was fashioned by my father from wood from a huge cherry tree that he, my brother David and I chopped down on my grandfather's farm in northeastern Ohio when I was five years old. David was two years older than me, so obviously, we really just watched as our dad chopped the tree down. He spent many of his retirement years down in his basement, listening to opera and making furniture out of that cherry tree. By the fire in my cozy study one evening in March, I was paging through the *Journal of Experimental Medicine* when I happened upon an article about the injection of interleukin-2 into the skin of individuals with leprosy (129). The investigators used low doses of interleukin-2 and found that it enhanced the generation of cytotoxic CD4+ T cells. I recalled that it was Ellis Reinherz who had first called attention to the fact that CD4+ T cells, which were usually thought of as "helper" T cells, can also become killers, especially if selected and grown long-term in media containing interleukin-2. Of the article's three authors, I knew the senior author, Zanvil Cohn. I fired off a short note to him, saying, "I enjoyed reading the study and thought that it was first rate." I soon received a reply that said, "Thanks, we think so too!" This was such a short and unusual reply that I wrote back a longer letter, telling him that I would like to come and visit him and his team to learn more about what they were doing. Thus began a collaboration that would take me half way around the world to Nepal, home to Mount Everest.

Zan Cohn was an established senior investigator at the Rockefeller University, one of the leading scientists studying phagocytic (eater) cells. I was intrigued that he was looking at my favorite molecule, interleukin-2, since my little molecule was a T cell growth factor, not a phagocytic cell growth factor. The Rockefeller University grew out of the Rockefeller Institute, which was established in 1901 by philanthropy from John D. Rockefeller at a time when bacteriology was ascendant, not long after the founding of the Pasteur Institute (1887) and the Robert Koch Institute in Berlin (1891). For the first half of the 20th century, "The Rock" was one of the only places in the United States where fundamental, basic research was done. It seems hard to believe now, but it was only after World War II, with the accelerated growth in federal government funding, that basic research began at the major American universities. As time went on during the 1950s and '60s, the Rockefeller Institute evolved, changing focus from the study of microbes to studies of the host response to the microbes. In 1965, the institute became the Rockefeller University, admitting graduate students. In the '70s and '80s, immunology became one of the major endeavors of the university.

I arranged to visit in the summer of 1989. Zan didn't ask me to give a seminar, but instead had me meet with the members of his lab. They had quite a factory devoted to all aspects of what Zan called the host-parasite interaction, a subject he had pioneered over his years at Rockefeller, which he had joined soon after Harvard Medical School in 1957. With the identification of interleukins as molecules, not merely mysterious activities, he had launched a series of studies, first *in vitro*, then *in vivo*, on the effects of selected interleukins, now that they were available in pure form and in great enough quantities to enable experimental approaches in humans to his fundamental questions. He was one of the few American scientists to take interleukins seriously. He had already investigated interferon-gamma, the interferon family member produced by lymphocytes, particularly T cells and natural killer cells. This was logical for him since interferon-gamma activates macrophages.

After a day chatting with the members of the Cohn lab, Zan invited me to drinks down in the Rockefeller bar on the ground floor of the administration building on campus and then to dinner with the entire lab, which was giving a going-away party for a member who was heading off to Seattle for another postdoc. The Rock was a surreal environment for the conduct of research with its quiet tree-lined campus of lab buildings in the middle of Manhattan on five blocks along the East River. The bar was equipped with lounge areas for comfortable discussions and pool tables to shoot the breeze. Although it was called a student and faculty club, it felt more like a student union where grad students and postdocs would congregate in the evenings. Later, at dinner when there was a lull in the conversation, Zan suddenly invited me to come to their lab on sabbatical. Of course, all went quiet as everyone awaited my reply. I can't remember exactly what I said, but I do believe I demurred, saying it was a kind and surprising offer and I would certainly consider it.

My foray down to the big city was successful on several levels and left me with a lot to think over. I admired Zan for his emphasis on human immunology and disease. In an attempt to bring the lab bench to the bedside, Zan had chosen leprosy as an infection that could be studied easily because the infection was primarily in the skin, which made it more accessible to biopsy and observation compared to an infection in internal organs. A few weeks later, I again heard from Zan with another invitation, this time to join his lab on a field trip. They were scheduled to travel to Nepal in November to investigate the administration of interleukin-2 into the skin of leprosy patients. Zan's group already had established that a single injection of interleukin-2 produced an accelerated delayed-type hypersensitivity reaction. It had the classic signs of inflammation although the inflammation was accelerated and visible 24 hours after injection instead of the usual 48 hours after antigen injection. But hypersensitivity was delayed compared to an immediate hypersensitivity reaction with inflammation visible in the skin after only a few minutes. This immediate reaction had already been shown by the Ishizakas to be mediated by antibodies and to be a sign of allergy. The delayed reaction had been established in the 1940s as the *in vivo* test for cellular immunity worked out by Merrill Chase, another Rockefeller scientist (130). He showed that it was possible to transfer this skin hypersensitivity from one experimental animal to another with cells, but not with plasma or sera. This demonstration separated cellular immunity from humoral (antibody) immunity. I thought the demonstration that interleukin-2 caused a delayed hypersensitivity response was highly significant. I pondered Zan's offer for a while and had to decide if it would be possible for me to leave my lab at that point and travel halfway around the world on a mission lasting two to three weeks. Carol Beadling was well into her search for interleukin-2-induced genes using the Pardee methods; Nina Zmuidzinas was working well with Harvey Mamon on *raf*; and Stefan was well ensconced with his memory project. I figured that it was now or never. I would join the Rock lab on this adventure.

We left JFK on the first weekend in November, the date of the annual New York Marathon. The city was energized with runners from around the globe. This heightened my feeling that I was off on a great adventure. Zan had left town earlier and we were to meet him in Nepal, we being Gilla Kaplan, one of the younger scientists from Zan's group, and two of her postdocs. Our flight was nonstop to Tokyo, then on to Hong Kong for an overnight stay, after that Bangkok and finally to Katmandu, the capital of Nepal. We were met at the airport by an Australian physician, Warwick Britton, one of the docs who manned a leprosarium outside the city, the Mission Anandaban Hospital. We climbed aboard a Land Rover with our carry-on luggage while the rest of our gear would arrive later. Driving through Katmandu, we passed a fetid pond occupied by a water buffalo and a woman washing her toddler's hair with the obviously contaminated water. There in a flash was my first understanding of the evolution and rigorous natural selection of our wonderful immune system. After all, it's only since the 1950s that we've only had widespread access to antibiotics, which extended the average human life expectancy from around 50 to almost 80 years. For humans to have survived in the midst of all these microbes, our immune system was the sole defense against microbial invasion we had. And it has done a marvelous job, for the most part. Over the centuries, endemic microbes as well as epidemics and pandemics have exerted a strong selective force on the effectiveness of our immune defenses to allow survival of the fittest.

After an hour's ride on unpaved roads, we reached our destination in the foothills of the Himalayas. We could no longer see Katmandu, as the leprosarium was on the backside of a large foothill. Our hosts later told us that the founders had felt that it wouldn't do to be able to actually see the leper hospital from the city and be reminded of its existence. Lepers were outcasts. On the first full day at the leprosarium just after sunrise, I took a hike to the top of the hill where I all of a sudden had a marvelous view of the snow-capped Himalayas on the horizon. It was a sight I never dreamed I would see.

In addition to Zan and Gilla Kaplan, who was originally from Israel, the Rockefeller team consisted of Gilla's two postdocs Jerry Hancock from the U.S. and Anthony Molloy from the U.K. Zan was 63 years old and had been investigating the host response to microbial infections for more than three decades. He impressed me as a benevolent, probing professor at the top of his game. From the beginning of our acquaintance, he did everything he could to help me. After a while together, I told him that I appreciated his style for the way he tried to help people, especially those who were younger and trying to climb the academic-scientific ladders. Even though I was 16 years his junior, I felt as if we were mutually supportive collaborators, answering the same questions with different approaches. I found his team immediately to my liking. I was used to working closely with a small group and being responsible for one of the techniques or assays. A few years younger than me, Gilla Kaplan was clearly the principal investigator on this project and especially kind in welcoming me to the team. She was a resolute scientist, focused on making sure that all went smoothly, and that Jerry Hancock and Anthony Molloy (and I) kept our noses to the grindstone.

Gilla and Zan had carefully planned the whole procedure. Having given just single doses of interleukin-2 to people without any adverse events, they would now treat the leprosy patients with twice daily injections into the skin at 12-hour intervals (9:00 a.m. and 9:00 p.m.) for eight days. Because Zan was interested in trying to understand how a delayed-type hypersensitivity (DTH) response occurred, he wanted to study any changes in both the immune system and the histopathology (microscopic appearance of cells and tissues) generated by interleukin-2, as well as the effects on the leprosy bacteria. Zan had already established collaborative contacts with experts in histopathology and microbiology of leprosy. Gilla's group was practiced in electron microscopy and immunological tests. I volunteered to examine the white blood cells by means of the flow cytometer to monitor changes in the subsets of lymphocytes

Dr. Britton assembled the volunteers. We injected the interleukin-2 solution into the skin on their backs. The morning shots were on one side of the back, at night on the other side. In November, with the rainy season over, the days were sunny and warm. We performed the injections outside on the roof of the hospital. The countryside was stunning with terraced rice paddies etched on the steep terrain. There were few roads to speak of, mostly footpaths meandering up and down the slopes, some of them quite steep. Given our morning and evening work hours, the afternoons were free for us to explore along the paths. The Nepalis divided the labor on the farms by gender. The men plowed the fields with oxen, the women harvested the rice. Since this was harvest season, most of the men were idle. We encountered them napping, lying on mats at the side of the path while the women toiled in the fields. This remembrance brings to mind the sweet smell that permeated the countryside, which I learned came from the rice wine the men drank.

Every so often we came upon one of the farmers' thatch-roofed mud huts. We were invited into some of these abodes. They had dirt floors and a single room occupied by entire families, together with various animal species. Seeing how these people lived, without basic sanitation, I understood why leprosy was endemic in this country unlike the developed world where public health sanitation measures had eradicated it. Still, I found it curious that only a few of the family members would become infected with the leprosy bacillus, which is similar to the bacterium that causes tuberculosis. I had learned that this was not a highly contagious bug, so it was perplexing why some individuals of a family became infected and some did not. Once infected, there were two clinical courses. For tuberculoid leprosy, the host's immune system attacked the bacteria, ultimately containing the organisms into a walled-off granuloma, the pathology term for a collection of macrophages that form a tubercle nodule under the skin. The other clinical course, lepromatous leprosy, was characterized by the lack of granuloma formation with many bacilli growing apparently unperturbed in the skin. The end result of the chronic infestation, especially of the skin, was damage to the sensory and pain nerve endings. People suffering from the infection lost their perception of pain and injury so that a characteristic picture was the loss of the terminal portions of fingers and toes. The faces of many leprosy sufferers were scarred and often their noses were missing. Once I had seen these manifestations in our study volunteers, it was easy to recognize the lepers on our trips to nearby villages.

While we were in Nepal, on November 8, 1989, we learned via the BBC over short-wave radio that the Berlin wall had fallen. My reaction was one of shock that this bastion of the cold war had been taken down by popular uprising. It was a harbinger of the collapse of the Soviet Union, which came two years later. Now more than 30 years later, as the French say, *plus ca change, plus la meme chose*. We are back to facing totalitarian Russia in the war in Ukraine.

The results of our study were significant (131). First, the administration of a low dose of interleukin-2 (10 micrograms) twice daily yielded no adverse events or discernible side effects. This was important, given Rosenberg's experience with high-dose interleukin-2 therapy. Second, there were readily detectable clinical and laboratory interleukin-2 effects on the immune system. Most of the subjects developed nontender enlarged lymph nodes under each arm, but they had no systemic symptoms. Third, with each injection, the sites showed progressively larger zones of delayed hypersensitivity with swelling, tenderness, redness and warmth to the touch. Fourth, like the experience with a single interleukin-2 injection, the reactions peaked after only 24 hours as compared with antigen-induced delayed hypersensitivity, like the TB skin test, which usually occurred after 48 hours. Presumably, the accelerated interleukin-2 reactions occurred because antigen-induced reactions required 24 hours to produce interleukin-2. This, in and of itself, was a revelation. The genesis of the delayed hypersensitivity reaction had been obscure. Now we realized that delayed hypersensitivity responses were due to the production of cytokines, especially interleukin-2, produced by T cells and natural killer cells at sites of antigen injection. Over the eight-day course of treatment, there were also changes observed in the concentration of white blood cells circulating in the blood: T cells, both helpers and killers, increased 2-2.5-fold, and natural killer cells increased six-fold. The effect of the interleukin-2 immunotherapy on the leprosy infection evolved over the ensuing months. Two months after the injections, there was an average five-fold reduction of bacteria in the skin at biopsy sites distant from the interleukin-2 injections. An identical result required a full year of chemotherapy with several drugs. Clearly, the interleukin-2 immunotherapy was effective in stimulating the immune system without toxicity, and it was effective as an antimicrobial.

This expedition had a profound effect on me. I realized a vast area of clinical research — trying to boost immune reactivity in situations where there might be a relative immunodeficiency, as in lepromatous leprosy — could very well be what I was looking for to expand the interests and expertise of my lab. Nowadays called translational medicine, this has become a favorite of funding agencies because it promises the translation of basic research findings to the bedside. And *bien sur* the pharmaceutical industry is clearly in favor of translational medicine.

Although it had been a lot of fun, I thought treating leprosy patients in Nepal (there were none in the U.S.) was a difficult way to go about conducting clinical research. Even though our initial results appeared promising, I knew it would be necessary to perform much longer trials, monitoring the subjects carefully and as thoroughly as possible. The need to spend months in Nepal, away from Hanover and my family, not to mention my lab, did not seem like a prudent option. Anyway, the leprosy studies had also already been claimed by Gilla and Zan, so I felt that avenue was not open. They felt that they had proven their hypothesis that leprosy was amenable to immunotherapy with interleukin-2, and Gilla subsequently went on to study TB in the states.

After the completion of our studies in Nepal, the group split up, each of us going our separate ways back to the States. I flew from Katmandu to Bangkok, then on to Hong Kong where I stayed overnight at the Mandarin Oriental Hotel. Quite an elegant place after our spartan existence in Nepal and a dramatic illustration of the first world vs. the third world. On the 13-hour flight back to New York, I had a lot of time to think about what we might do next. With the onset of AIDS in 1983, I felt certain I had to get involved in some way, especially when the receptor for HIV was found to be the CD4 molecule expressed on the surface of helper T cells. Yet as the 1980s got underway, I had elected to stay away from this field for several reasons. First, those suffering from AIDS were very sick and as time went on with deaths accumulating, the patients, their families and friends became increasingly angry and desperate. The physicians on the front lines were also distraught because their patients were suffering and dying so terribly. As to treatment, the antimicrobials available were ineffective in combating the many opportunistic infections in the face of the profound immunodeficiency. This was a dramatic example of what happens when the immune system fails. Finally, the scientists who dove into this field were also frantic, as they recognized that one could become famous overnight by contributing to solving this problem with the world focused on it. I could not think of a way that I could add to the research on a basic level, in which many others, most of them virologists, were already engaged. As an example, I received a call at home on a Sunday afternoon in the early '80s from Bob Gallo who informed me that I needn't worry about him working on interleukin-2 anymore. He always thought he discovered interleukin-2, but I knew his lab had only rediscovered that lymphocyte-conditioned media contains lymphocyte growth factors. He said was going to devote his entire lab full time to work on AIDS and HIV. I thought that was great because at least I did not have to compete with his machine. As time went on, and as a result of my Nepal experience, the idea of treating AIDS patients with interleukin-2, in hopes of boosting their immunity, just as it had seemed to do for the leprosy patients, became more and more attractive. This was also the time when I was collaborating with Jerry Ritz and Michael Caligiuri in their clinical trials using interleukin-2 to try to boost immune reactivity in bone-marrow transplant recipients. In those studies, we found interleukin-2 could be safely administered intravenously for as long as three months in daily continuous IV infusions.

As a result of my work with both Zan Cohn and Jerry Ritz who used low doses of interleukin-2 without toxicity, I formulated a plan to boost the immune reactivity of AIDS patients. In 1990 and 1991, the HIV epidemic was raging, thousands were dying from multiple microbial infections as a result of their immunodeficiencies. I almost felt obligated to try low-dose interleukin-2 therapy in these patients. This presented a problem. At our hospital, there were few AIDS patients. It is difficult to conduct clinical research without experimental volunteers. I went to the dean of Dartmouth Med School to inquire about what we might be able to do to get more patients. He was not encouraging. In fact, he was downright discouraging. "AIDS is a big city disease," he told me, and the patients were mostly "undesirables," being homosexuals or IV drug abusers.

He did not want a lot of these patients to come to our hospital in the pristine Upper Valley.

With this as a background, I took Zan Cohn up on his offer to come to his lab on a sabbatical. In 1990, I worked directly with Gilla Kaplan who was investigating thalidomide, one of the only therapies effective in reducing the often-severe inflammatory skin reactions seen in leprosy patients. In the 1950s, this compound had been given to pregnant women as an antinausea agent and was found to cause serious birth defects. Luckily, this time we found that thalidomide reduced the production of tumor necrosis factor, and our report remains one of my most frequently cited papers (132).

While collaborating with Zan and Gilla, I was approached by Ralph Nachman who was the newly appointed chairman of the Department of Medicine at Cornell University Medical College in New York City, which was literally next door to Rockefeller. I figured that Zan Cohn turned him onto me. Ralph was looking for a chief of the Division of Allergy and Immunology. I met with him and shared my vision of creating a new subspecialty in medicine, medical immunology, similar to medical oncology, which had been built upon the development of specialized (and toxic) chemotherapy of cancer. I felt we were on the threshold of a new era, of using the new cytokine molecules in the clinic to boost immune reactivity for diseases where there was a clear deficiency of the immune system, AIDS being the most obvious and pressing need. Ralph was receptive to this idea. We began negotiating the particulars of what a job at Cornell would entail, and I started making plans to move to New York City.

EPILOGUE

At the end of two decades of fundamental research into how normal cells make the decision to divide, which led us to uncover the hormonal control of the immune response, my career had come to a fork in the road. To go further with trying to understand the molecular mechanisms of cell growth, both normal and malignant, it was clear that two areas of basic science had to be attacked. One was the biochemistry of receptor-mediated signal transduction; the second was the molecular biology driving cell proliferation.

After a decade of oncogene research by others, I knew this field was hamstrung by the lack of assays to identify how and where the oncogenes function inside the cell. The prospect of using our interleukin-2/T cell proliferation model afforded us the opportunity to move to molecular causes of both normal cell growth and malignancy. Our initial discoveries in this endeavor pointed to two branches: one was signal transduction, the immediate cytoplasmic biochemical reactions activated by the interleukin-2 receptor; the other was the nuclear molecular control of DNA replication and mitosis. Of these two, signal transduction was obviously going to be very competitive. It was already evident that some of the best labs in the world were attacking it vigorously. I knew we just needed to follow our noses into signal transduction and chart the course of signals emanating from the interleukin-2 receptor in the initial seconds and minutes into the cell interior. However, I also knew that this course was fraught with a multitude of unknowns, and it could very well become a morass with many branched paths. In addition, the biochemical assays available for this pursuit were for the most part qualitative, not quantitative. Based on my principles of investigation, this was a significant deterrent.

Equally beckoning was the lure of using interleukin-2 in the clinic to try to boost immunoreactivity. In trying to choose between these paths, the prospect of using interleukin-2 and perhaps other cytokines in the clinic to stimulate immune responses appeared to have great potential, because immunostimulation as a new form of therapy had not yet been developed to become commonplace in the annals of medicine as had immunosuppression. It was clear that there was a need for immunostimulatory therapy, especially for cancer, as exemplified by the efforts in years past of William Coley, Georges Mathé and Steve Rosenberg. But immunopathology, which led me from medicine to science in the first place, was a significant obstacle to stimulatory immunotherapy. We had made significant progress in understanding the exact molecular mechanisms responsible for immunopathology. It was due to inflammation, which we had realized was due to the pro-inflammatory cytokines. Based on this understanding, I felt that we could develop a rational method for immunostimulation with interleukin-2 that avoided toxicity, and clinical research was beckoning. The key to the manipulation of the immune response was our novel understanding of its hormonal characteristics. I realized that whole new therapeutic modalities should be forthcoming by manipulating these new hormones, their receptors and their signaling pathways.

Thus, going forward, I elected to focus the lab onto two divergent paths. One would be the molecular biology of cellular proliferation and hopefully eventually oncogenesis; the other the clinical research of immunostimulatory therapy for HIV/AIDS.

I figured that would keep me busy for the foreseeable future.

ACKNOWLEDGEMENTS

As I was immersed in my science for so many years, striving to "push back the frontiers" trying to be first to discover new knowledge, I would often think back to my earliest years in school when I labored to learn the myriad of facts presented to me in all of the textbooks. Most of the time, the facts were presented as "written in stone", appearing to have been established by some ancient authority. Seldom were the facts attributed to people. Science seemed to be set apart from the humanities, which were all about the people. I have also been dismayed that in our popular culture, science as a profession or as a lifelong endeavor is seldom mentioned as a potential pursuit for young people to consider. Rather, politics, entertainment, athletics, and business are constantly in the news.

When I set out to write this memoir of my first twenty years of my career, I wanted to tell my story in hopes that young people might be able to gain some insight into what science is like and what the people are like who are driven to it for a career. Science is perhaps the quintessential creative endeavor, in that science is the human process by which new knowledge is created. Thus, people create science by using the scientific method, and I have tried to highlight that fact by detailing each contribution that moved our quest forward. Hopefully there will be a scientist described in this volume who will catch the imagination of a new young person to serve as a model.

In thinking about the many people who have helped me in my quest and who thus have contributed to this volume, I must begin with mentioning my parents. My father taught me in a fascinating way about all of the wonders of nature throughout my childhood.

He led me by the hand through the forest and opened my eyes to all of the living things in our environment, both flora and fauna, thereby quenching my incessant curiosity. My mother instilled in me a sense of self confidence by repeating over and again that "you can do anything that you want to do, as long as you set your mind to it".

I could never have written this book without the support of my high school sweetheart, Lynn. We essentially raised each other from childhood to adulthood. She has experienced along with me the highest of highs as well as the lowest of lows.

I owe my love of writing to Maurice Landy, an editor for Academic Press, who critiqued my manuscripts through the first years of the discoveries detailed in this volume. Whenever I think of him, I hear his voice saying that "scientific writing must be concise, tightly integrated, and persuasive".

I have always said that if you want to learn anything new, it helps immensely to have a good teacher. In the case of my artistic efforts, which I started 15 years ago, my work improved progressively when I started with Nancy Goff and delved into pastels, which are included in this book.

Once I had a reasonable first draft in hand, I contacted an old friend, Gary Taubes for advice on how to proceed to publishing. Gary is a scientific writer focused on nutrition and how our western diet has led to the obesity epidemic, which is now worldwide. Gary recommended that I contact Marion Roach Smith, an old friend of his who specializes in helping people with writing memoirs. Marion was critical in the early stages with advice on developmental and structural aspects, then referred me to Rob Brill, who spent a year with me, examining the use of so many scientific and medical terms that are in common use but often mean different things to the scientific aficionados.

During the writing of this book, the coronavirus pandemic occurred. This provided for me a living example of how important it is that immunology is understood by everyone. This spurred me on to write about the excitement of discovering how the immune system is regulated and why vaccination is so important and how inflammation ultimately can lead to severe disease and death. Now, more than ever we need young people to enter into immunology and continue the quest for new knowledge.

REFERENCES

1. LeBauer E, Smith K, and Greenberger N. Pancreatic insufficiency and vitamin B12 malabsorption. *Arch Int Med.* 1968;122:423-5.

2. Nowell PC. Phytohemagglutinin: An initiator of mitosis in cultures of normal human leukocytes. *Cancer Research.* 1960;20:462-8.

3. Smith K, Chess L, and Mardiney M. The characteristics of lymphocyte tritiated thymidine incorporation in response to mumps virus. *Cell Immunol.* 1972;5:597-603.

4. Burnet FM. A modification of Jerne's theory of antibody production using the concept of clonal selection. *Aust J Sci.* 1957;20:67-77.

5. Coley W. The treatment of malignant tumors by repeated inoculations of erysipalas: with a report of ten original cases. *Am J Med Sci.* 1893;10:487-511.

6. Freund J, and McDermott K. Sensitization to horse serum by means of adjuvants. *Proc Soc, Exp Biol Med NY.* 1942;49:548-33.

7. Oppenheim J, Leventhal B, and Hersh E. The transformation of column-purified lymphocytes with nonspecific and specific antigenic stimuli. *J Immunol.* 1968;101(2):262-70.

8. Chess L, Schmuckler M, Smith K, and Mardiney M. The effect of synthetic polyribonucleotides on immunologically induced tritiated thymidine incorporation. *Transplantation.* 1972;14:748-56.

9. McIntyre O, and Ebaugh F. The effect of phytohemagglutinin on leukocyte cultures as measured by P32 incorporation in the DNA, RNA, and acid soluble fractions. *Blood.* 1962;19:443-53.

10. Issacs A, Lindeman J, and Valentine RC. Virus interference. II. Some properties of interferon. *Proc Roy Soc.* 1957;B 147:268-73.

11. Cornell C, Smith K, Cornwell G, Burke G, and McIntyre O. Systemic effects of intravenous polyriboinosinic-polyribocytidylic acid in man. *J Natl Cancer Inst.* 1976;57:1211-6.

12. Smith K, Cornwell G, and McIntyre O. *Proceedings of the Eighth Leukocyte Culture Conference: Lymphocyte Recognition and Effector Mechanisms*. Academic Press, Inc.; 1974:101-9.

13. Mathe G, Jammet H, Pendic B, Schwarzenberg L, Duplan J-F, Maupin B, et al. Transfusions et greffes di moelle osseuse homologue chez des humains irradies a haut dose accidentellement. *Rev France Etudes Cli, et Biol.* 1959;4:226-38.

14. Mathe G, Amiel J, Schwarzenberg L, Cattan A, and Schneider M. Haematopoietic chimera in man after allogeneic (homologous) bone-marrow transplantation. *Brit Med J.* 1963;5373:1633-5.

15. Mathe G. Lecon Inaugurale: Faculte de Medecine de Paris, Chaire de Cancerologie Experimentale. *La Presse Medicale.* 1967;75:841-6.

16. Mathe G, Amiel J, Schwarzenberg L, Schneider M, Cattan A, Schlumberger J, et al. Active immunotherapy for acute lymphoblastic leukemia. *The Lancet.* 1969;293(7597):697-9.

17. Calmette A, Guerin C, Weill-Halle B, and al. e. Essais d'immunisation contre l'infection tuberculeuse. *Bulliten de l'Academie de Medicine.* 1924;91:787-96.

18. Freund J, and McDermott K. Sensitization to horse serum by means of adjuvants. *Proc Soc Exp Biol Med.* 1942;49:548-53.

19. Wilson E. *Sociobiology.* Cambridge, MA: The Bellnap Press of Harvard University Press; 1975.

20. Pasteur L. Methode pour prevenir la rage apres morsure. *Compte Rendus de l'Acad Sci.* 1885;101:765-74.

21. Mathe G, and Amiel J. The roles of adoptive and active forms of iummunotherapy in the cure of children suffering from acute lymphoid leukemia: a) underestimation of active ikmmunotherapy benefit, b) its immunogenetic indications to select sensitive patients, hence prevent chemotherapy's late effects. . *Biomed Pharmacother.* 2001;55:531-42.

22. Bach F, Alter B, Solliday S, Zoschke D, and Janis M. Lymphocyte reactivity in vitro II. Soluble reconstituting factor permitting response of purified lymphocytes. *Cell Immunol.* 1970;1:219-27.

23. Gery I, Gershon RK, and Waksman B. Potentiation of the T-lymphocyte response to mitogens I. The responding cell. *J Exp Med.* 1972;136:128-42.

24. Gery I, and Waksman BH. Potentiation of the T-lymphocyte response to mitogens: The cellular source of potentiating mediators. *J Exp Med.* 1972;136:143-55.

25. Mathe G, Kamel M, Dezfulian M, Halle-Pannenko O, and Bourut C. An experimental screening for "systemic adjuvants of immunity" Application in cancer immunotherapy. *Cancer Res.* 1973;33:1987-97.

26. Fredrickson TN, Smith KA, Cornell CJ, Jasmin C, and McIntyre OR. The interaction of erythropoietin with fetal liver cells I. Measurement of proliferation by tritiated thymidine incorporation. *Exp Hematol.* 1977;5:254-65.

27. Smith K, Fredrickson T, Mobraaten L, and DeMaeyer E. The interaction of erythropoietin with fetal liver cells. II. Inhibition of the erythropoietin effect by interferon. *Exp Hemat.* 1977;5:333-40.

28. Wagner H, and Feldmann M. Cell-mediated immune response in vitro: I. A new in vitro system for the generation of cell-mediated cytotoxic activity. . *Cell Immunol.* 1972;3:405-20.

29. Schmeck HJ. Litton to run cancer research lab. https://www.nytimes.com/1972/06/25/archives/litton-to-run-cancer-research-lab.html.

30. Gillis S, and Smith K. In vitro generation of tumor-specific cytotoxic lymphocytes. Secondary allogeneic mixed tumor lymphocyte culture of normal murine spleen cells. *J Exp Med.* 1977;146:468-82.

31. Cline M, and Golde D. Production of colony-stimulating activity by human lymphocytes. *Nature.* 1974;248:703-4.

32. Wheelock E. Interferon-like virus inhibitor induced in human leukocytes by phytohemagglutinin. *Science.* 1965;149:310-1.

33. Kasakura S, and Lowenstein L. A factor stimulating DNA synthesis derived from the medium of leukocyte cultures. *Nature.* 1965;208:794-5.

34. Gordon J, and MacLean LD. A lymphocyte-stimulating factor produced in vitro. *Nature.* 1965;208:795-6.

35. Smith K. *Molecular Immunity: A Chronology of 60 Years of Discovery.* Singapore: World Scientific Publishing Co. Pte. Ltd.; 2018.

36. Morgan DA, Ruscetti FW, and Gallo R. Selective in vitro growth of T lymphocytes from normal human bone marrows. *Science.* 1976;193(4257):1007-8.

37. Gillis S, and Smith KA. Long term culture of tumour-specific cytotoxic T cells. *Nature.* 1977;268(5616):154-6.

38. Medawar P. *Advice to a Young Scientist.* New York, NY: Basic Books; 1979.

39. Kohler G, and Milstein C. Continuous culture of fused cells secreting antibody of predefined specificity. *Nature.* 1975;256:495-9.

40. Baker PE, Gillis S, and Smith KA. Monoclonal cytolytic T-cell lines. *J Exp Med.* 1979;149(1):273-8.

41. Baker PE, Gillis S, Ferm MM, and Smith KA. The effect of T cell growth factor on the generation of cytolytic T cells. *J Immunol.* 1978;121(6):2168-73.

42. Smith K, Gilbride K, and Favata M. Lymphocyte activating factor promotes T cell growth factor production by cloned murine lymphoma cells. *Nature.* 1980;287:853-5.

43. Smith KA, Lachman LB, Oppenheim JJ, and Favata MF. The functional relationship of the interleukins. *J Exp Med.* 1980;151(6):1551-6.

44. Smith KA. The interleukin 2 receptor. *Annu Rev Cell Biol.* 1989;5:397-425.

45. Munck A, and Brinck-Johnson T. Specific and nonspecific physicochemical interactions of glucocorticoids and related steroids with rat thymus cells in vitro. *J Biol Chem.* 1968;243:5556-65.

46. Smith KA, Crabtree GR, Kennedy SJ, and Munck AU. Glucocorticoid receptors and glucocorticoid sensitivity of mitogen stimulated and unstimulated human lymphocytes. *Nature.* 1977;267(5611):523-6.

47. Nowell PC. Inhibition of human leukocyte mitosis by prednisolone in vitro. *Cancer Research.* 1961;21:1518-23.

48. Gillis S, Crabtree GR, and Smith KA. Glucocorticoid-induced inhibition of T cell growth factor production. I. The effect on mitogen-induced lymphocyte proliferation. *J Immunol.* 1979;123(4):1624-31.

49. Gillis S, Crabtree GR, and Smith KA. Glucocorticoid-induced inhibition of T cell growth factor production. II. The effect on the in vitro generation of cytolytic T cells. *J Immunol.* 1979;123(4):1632-8.

50. Bunjies D, Hardt C, Rollinghoff M, and Wagner H. Cyclosporin A mediates immunosuppression of primary cytotoxic T cell responses by impairing the release of interleukin-1 and interleukin-2. *Eur J Immunol.* 1981;11:657-61.

51. Kuo C, Chung J, Fiorentino D, Flanagan W, Blenis J, and Crabtree G. Rapamycin selectively inhibits interleukin-2 activation of p70 S6 kinase. *Nature* 1992;358:70-3.

52. Robb RJ, and Smith KA. Heterogeneity of human T-cell growth factor(s) due to variable glycosylation. *Mol Immunol.* 1981;18(12):1087-94.

53. Robb RJ, Munck A, and Smith KA. T cell growth factor receptors: Quantitation, specificity, and biological relevance. *J Exp Med.* 1981;154(5):1455-74.

54. Kaplan J, Tilton J, and Peterson WJ. Identification of T cell lymphoma tumor antigens on human T cell lines. *Am J, Hematol.* 1975;1:219-27.

55. Gillis S, and Watson J. Biochemical and biological characterization of lymphocyte regulatory molecules. *J Exp Med.* 1980;152:1709-19.

56. Julius M, Masuda T, and Herzenberg L. Demonstration that antigen-binding cells are precursors of antibody-producing cells after purification with a fluorescence-activated cell sorter. *Proc Natl Acad Sci USA.* 1972;69:1934-8.

57. Leonard WJ, Depper JM, Uchiyama T, Smith KA, Waldmann TA, and Greene WC. A monoclonal antibody that appears to recognize the receptor for human T-cell growth factor; partial characterization of the receptor. *Nature.* 1982;300(5889):267-9.

58. Cantrell DA, and Smith KA. The interleukin-2 T-cell system: a new cell growth model. *Science.* 1984;224(4655):1312-6.

59. Larsson E, Iscove N, and Coutinho A. Two distinct factors are required for T cell growth. *Nature.* 1980;283:664-6.

60. Meuer SC, Fitzgerald KA, Hussey RE, Hodgdon JC, Schlossman S, and Reinherz EL. Clonotypic structures involved in antigen-specific human T cell function. Relationship to the T3 molecular comlex. *J Exp Med.* 1983;157:705-19.

61. Meuer S, Cooper D, Hodgdon J, Hussey R, Fitzgerald K, Schlossman S, et al. Identification of the receptor for antigen and major histocompatability complex on human inducer T lymphocytes. *Science.* 1983;222:1239-42.

62. Meuer SC, Hodgdon JC, Hussey RE, Protentis JP, Schlossman S, and Reinherz EL. Antigen-like effects of monoclonal antibodies directed at receptors on human T cell clones. *JExp Med.* 1983;158:988-93.

63. Allison J, McIntyre B, and Bloch D. Tumor-specific antigen of murine T-lymphoma defined with monoclonal antibody. *J Immunol.* 1982;129(5): p2293-300.

64. Haskins K, Kubo R, Pigeon M, White J, Kappler J, and Marrack P. The major histocompatability compex restricted antigen receptor in T cells. I. Isolation with a monoclonal antibody. *J Exp Med.* 1983;157:1149 58.

65. Hedrick S, Cohen D, Nielson E, and Davis MM. Isolation of cDNA clones encoding T cell-specific memebrane associated proteins. *Nature.* 1984;308:149-55.

66. Yanagi Y, Yoshikai Y, Leggett K, Clark S, Aleksander I, and Mak T. A human T cell-specific cDNA clone encodes a protein having extensive homology to immunoglobulin chains. *Nature* 1984;308:145-9.

67. Acuto O, Fabbi M, Smart J, Poole C, Protentis J, Royer H, et al. Purification and NH2-terminal amino acid sequencing of the beta subunit of a human T cell antigen receptor. *Proc Natl Acad Sci USA.* 1984;81:3851-5.

68. Meuer SC, Hussey RE, Cantrell DA, Hodgen JC, Schlossman SF, Smith KA, et al. Triggering the T3-Ti antigen-receptor complex results in clonal T cell proliferation through an interleukin 2-dependent autocrine pathway. *PNAS.* 1984;81:1509-13.

69. Smith KA. Interleukin-2: inception, impact, and implications. *Science.* 1988;240(4856):1169-76.

70. Gullberg M, and Smith KA. Regulation of T cell autocrine growth. T4+ cells become refractory to interleukin 2. *J Exp Med.* 1986;163(2):270-84.

71. Taniguchi T, Matsui H, Fujita T, Takaoka C, Kashima N, Yoshimoto R, et al. Structure and expression of a cloned cDNA for human interleukin-2. *Nature.* 1983;302(5906):305-10.

72. Smith KA, Favata MF, and Oroszlan S. Production and characterization of monoclonal antibodies to human interleukin 2: strategy and tactics. *J Immunol.* 1983;131(4):1808-15.

73. Budd R, and Smith KA. Interleukin 2 monoclonal antibody affinity adsorption: the critical role of binding kinetics for optimal immunoadsorption. *The Journal of Immunologic Methods*. 1986;95:237-48.

74. Budd R, and Smith KA. Interleukin 2 immunoassay using monoclonal antibodies. *Biotechnology*. 1986;4:983-6.

75. Cohen F, Kosen P, Kuntz I, Epstein L, Ciardelli T, and Smith K. Structure-activity studies of interleukin-2. *Science*. 1986;234:349-52.

76. Landgraf B, Cohen F, Smith K, Gadski R, and Ciardelli T. Structural significance of the C-terminal amphiphilic helix of interleukin-2. *J Biol Chem*. 1989;254:816-22.

77. Brandhuber BJ, Boone T, Kenney WC, and McKay DB. Three-dimensional structure of interleukin 2. *Science*. 1987;238:1707-9.

78. Stauber D, Debler E, Horton P, Smith K, and Wilson I. Crystal structure of the interleukin-2 signaling complex: paradigm for a heterotrimeric cytokine receptor. *PNAS*. 2006;103:2788-93.

79. Uchiyama T, Broder S, and Waldmann TA. A monoclonal antibody (anti-Tac) reactive with activated and functionally mature human T cells. I. Production of anti-Tac monoclonal antibody and distribution of Tac (+) cells. *Journal of Immunology*. 1981;126:1393-7.

80. Smith KA, and Cantrell DA. Interleukin 2 regulates its own receptors. *Proc Natl Acad Sci U S A*. 1985;82(3):864-8.

81. Rabin H, Hopkins R, Ruscetti F, Neubauer R, Brown R, and Kawakami T. Spontaneous release of a factor with properties of T cell growth factor from a continuous line of primate tumor cells. . *J Immunol*. 1981;127:1852-6.

82. Smith K. T cell growth factor and glucocorticoids: Opposing regulatory hormones in neoplastic cell growth. *Immunobiology*. 1982;161:157-73.

83. Kato K, and Smith K. Tac antigen forms disulfide-linked homodimers. *Biochemistry*. 1987;26:5359-64.

84. Teshigawara K, Wang HM, Kato K, and Smith KA. Interleukin 2 high-affinity receptor expression requires two distinct binding proteins. *J Exp Med*. 1987;165(1):223-38.

85. Sharon M, Klausner RD, Cullen BR, Chizzonite R, and Leonard WJ. Novel interleukin-2 receptor subunit detected by cross-linking under high affinity conditions. *Science.* 1986;234:859-63.

86. Tsudo M, Kozak RW, Goldman CK, and Waldmann TA. Demonstration of a non-Tac peptide that binds interleukin 2: A potential participant in a multichain interleukin 2 receptor complex. *Proceedings of the National Academy of Sciences USA.* 1986;83:9694-8.

87. Wang HM, and Smith KA. The interleukin 2 receptor. Functional consequences of its bimolecular structure. *J Exp Med.* 1987;166(4):1055-69.

88. Sutter A, Riopelle R, Harris-Warrick R, and Shooter E. Nerve growth factor recceptors: Characterization of two distinct classes of binding sites on chick embryo snensory ganglia cells. *J Biol Chem.* 1979;254:5972-82.

89. Temin H. Nature of the provirus of Rous sarcoma. *Nat Can Inst Monograph.* 1964;17:557-70.

90. Temin H. The protovirus hypothesis: Speculations on the significance of RNA-directed DNA synthesis for normal development and for carcinogenesis. *J Nat Can Inst.* 1971;46(2):III-VII.

91. Shih C, Shilo B-Z, Goldfarb MP, Dannenberg A, and Weinberg RA. Passage of Phenotypes of Chemically Transformed Cells Via Transfection of DNA and Chromatin. *PNAS.* 1979;76(11):5714-8.

92. Cooper G, Okenquist S, and Silverman L. Transforming activity of DNA of chemically transformed and normal cells. *Nature.* 1980;284:418-21.

93. Stern JB, and Smith KA. Interleukin-2 induction of T-cell G1 progression and c-myb expression. *Science.* 1986;233(4760):203-6.

94. Stehelin D, Varmus H, Bishop J, and Vogt P. DNA related to the transforming gene(s) of avain sarcoma viruses present in normal avian DNA. *Nature.* 1976;260:170-3.

95. Joaquin M, and Watson R. Cell cycle regulation by the B-Myb transcription factor. *Cell Mol Life Sci.* 2003;60:2389-401.

96. Ramsay R, and Gonda T. MYB function in normal and cancer cells. *Nature Rev Cancer.* 2008;8:523-34.

97. Rapp U, Goldsborough M, Mark G, Bonner T, Groffen J, Reynolds FJ, et al. Structure and biologicl activity of v-raf, a unique oncogene transduced by a retrovirus.. *Proc Natl Acad Sci USA.* 1983;80:4218-22.

98. Zmuidzinas A, Mamon HJ, Roberts TM, and Smith KA. Interleukin 2-triggered Raf-1 expression, phosphorylation, and associated kinase activity increase through G1 and S in CD3-stimulated primary human T cells. *Mol Cell Bio.* 1991;11:2794-803.

99. Woodford T, Schlegel R, and Pardee A. Selective isolation of newly synthesized mamaliian mRNA after in vivo labeling with 4-thiouridine or 6-thioguanine. *Anal Biochem.* 1988;171:166-72.

100. Beadling C, Johnson KW, and Smith KA. Isolation of interleukin 2-induced immediate-early genes. *Proc Natl Acad Sci U S A.* 1993;90(7):2719-23.

101. Grimm E, Mazumder A, Zhang H, and Rosenberg S. Lymphokine-activated killer cell phenomenon. Lysis of natural killer resistant fresh solid tumor cells by interleukin 2-actgivated autologous human peripheral lymphocytes. *J Exp Med.* 1982;155:1823-41.

102. Phillips J, and Lanier L. Dissection of the lymphokine-activated killer cell phenomenon. Relative contribution of peripheral blood natural killer cells and T lymphocytes to cytolysis. *J Exp Med.* 1986;164:814-25.

103. Kiessling R, Klein E, Pross H, and Wigzell H. Natural killer cells in the mouse. II. Cytotoxic cells with specificity for mouse Moloney leukemia cells. Characteristics of the killer cell. *Eur J Immunol.* 1975;5:117-22.

104. Herberman RB, Nunn ME, Holden HT, and Lavrin DH. Natural cytotoxic reactivity of mouse lymphoid cells against syngeneic and allogeneic tumors. II. Characterization of effector cells. *Int J Cancer.* 1975;16:230-9.

105. Rosenberg SA, Lotze MT, Muul LM, Chang AE, Avis FP, Leitman S, et al. A progress report on the treatment of 157 patients with advanced cancer using lymphokine-activated killer cells and interleukin-2 or high-dose interleukin-2 alone. *N Engl J Med.* 1987;316(15):889-97.

106. Rosenberg S, Lotze M, Yang J, Aebersold P, Linehan W, Seipp C, et al. Experience with the use of high-dose interleukin-2 in the treatment of 652 cancer patients. *Ann Surg.* 1989;210:474-85.

107. Rosenberg S, Lotze M, Yang J, Topalian S, Chang A, Schwartzentruber D, et al. Prospective randomized trial of high-dose interleukin-2 alone or in conjunction with lymphokine-activated killer cells for the treatment of patients with advanced cancer. *J Natl Cancer Inst.* 1993;85:622-32.

108. Granger G, and Williams T. lymphocyte cytotoxicity in vitro: activation and release of a cytotoxic factor. *Nature.* 1968;218:1253-4.

109. Ruddle N, and Waksman B. Cytotoxicity mediated by soluble antigen and lymphocytes in delayed hypersensitivity. III. Analysis of mechanism. *J Exp Med.* 1968;128:1267-79.

110. Carswell E, Old L, Kassel R, Green S, Fiore N, and Williamson B. An endotoxin-induced serum factor that causes necrosis of tumors. *Proc Natl Acad Sci U S A.* 1975;72(9): p3666-70.

111. Blick M, Sherwin S, Rosenblum M, and Guttermen J. Phase I study of recombinant tumor necrosis factor in cancer patients. *Cancer Res.* 1987;47:2986-9.

112. Creaven P, Plager J, Dupere S, Huben R, Takita H, and Mittlemen A. Phase I clinical trial of recombinant human tumor necrosis factor. *Cancer Chemother Pharmacol.* 1987;20:137-44.

113. Spriggs D, Sherman M, Michie H, Arthur K, Imamura K, Wilmore D, et al. Recombinant human tumor necrosis factor administered as a 24-hour continuous infusion. A phase I and pharmacologic study. . *J Natl Ca Institute.* 1988;80:1039-44.

114. Beutler B, and Cerami A. The biology of cachectin/TNF--a primary mediator of the host response. *Annu Rev Immunol.* 1989;7:625-55.

115. Beutler B, and Cerami A. Cachectin and tumour necrosis factor as two sides of the same biological coin. *Nature.* 1986;320(6063): p584-8.

116. Michie H, Manogue K, Spriggs D, Revhaug A, O"Dwyer S, Dinarello C, et al. Detection of circulating tumor necrosis factor after endotoxin administration. *New Eng J Med.* 1988;318:1481-6.

117. Gemlo B, Palladino M, Jaffe H, Espevik T, and Rayner A. Circulating cytokines in patients with metastatic cancer treated with recombinant interleukin 2 and lymphokine-activated killer cells. *Cancer Res.* 1988;48:5864-7.

118. Mier J, Vachino G, Van der Meer J, Numerof R, Adams S, Cannon J, et al. Induction of circulating tumor necrosis factor (TNFa) as the mechanism for the febrile response to interleukin-2 administration in cancer patients. *J Clin Immunol.* 1988;8:426-36.

119. Caligiuri MA, Zmuidzinas A, Manley TJ, Levine H, Smith KA, and Ritz J. Functional consequences of interleukin 2 receptor expression on resting human lymphocytes. Identification of a novel natural killer cell subset with high affinity receptors. *J Exp Med.* 1990;171(5):1509-26.

120. Caligiuri MA, Murray C, Soiffer RJ, Klumpp TR, Seiden M, Cochran K, et al. Extended continuous infusion low-dose recombinant interleukin-2 in advanced cancer: prolonged immunomodulation without significant toxicity. *J Clin Oncol.* 1991;9(12):2110-9.

121. Takeshita T, Ohtani K, Asao H, Kumaki S, Nakamura M, and Sugamura K. An associated molecule, p64, with IL-2 receptor beta chain. Its possible involvement in the formation of the functional intermediate-affinity IL-2 receptor complex. *J Immunol.* 1992;148(7):2154-8.

122. Takeshita T, Asao H, Ohtani K, Ishii N, Kumaki S, Tanaka N, et al. Cloning of the gamma chain of the human IL-2 receptor. *Science.* 1992;257(5068):379-82.

123. Sana T, Wu Z, Smith K, and Ciardelli T. Expression and ligand binding characterization of the beta subunit (p75) ectodomain of the interleukin-2 receptor. *Biochem* 1994;33:5838-45.

124. Sugamura K, Asao H, Kondo M, Tanaka N, Ishii N, Nakamura M, et al. The common gamma-chain for multiple cytokine receptors. *Adv Immunol.* 1995;59:225-77.

125. Caligiuri MA, Murray C, Robertson MJ, Wang E, Cochran K, Cameron C, et al. Selective modulation of human natural killer cells in vivo after prolonged infusion of low dose recombinant interleukin 2. *J Clin Invest.* 1993;91(1):123-32.

126. Grabstein K, Eisenman J, Shanebeck K, Rauch C, Srinivasan S, Fung V, et al. Cloning of a T cell growth factor that interacts with the beta chain of the interleukin-2 receptor. *Science.* 1994;264(5161): p965-8.

127. Carson W, Giri J, Lindeman M, Linett M, Ahdieh M, Paxton R, et al. Interleukin (IL) 15 is a novel cytokine that activates human natural killer cells via components of the interleukin-2 receptor. *J Exp Med.* 1994;180:1395-403.

128. Handa K, Suzuki R, Matsui H, Shimizu Y, and Kumagai K. Natural killer (NK) cells as a responder to interleukin 2 (interleukin-2) interferon gamma production. *J Immunol.* 1983;130:988-92.

129. Hancock G, Cohn Z, and Kaplan G. The generation of antigen-specific, major histocompatibility complex-restricted cytotoxic T lymphocytes of the CD4+ phenotype. Enhancement by the cutaneous administration of interleukin 2. *J Exp Med.* 1989;169:909-19.

130. Chase M. The cellular transfer of cutaneous hypersensitivity to tuberculin. *Proc Soc Exp Biol Med.* 1945;59:134-5.

131. Kaplan G, Britton WJ, Hancock GE, Theuvenet WJ, Smith KA, Job CK, et al. The systemic influence of recombinant interleukin 2 on the manifestations of lepromatous leprosy. *J Exp Med.* 1991;173(4):993-1006.

132. Moreira A, Sampaio E, Zmiudzinas A, Frindt P, Smith K, and Kaplan G. Thalidomide exerts its inhibitory action on tumor necrosis factor alpha by enhancing mRNA degradation. *J Exp Med.* 1993;177:1675-80.

NAME INDEX

SUBJECT INDEX

Made in the USA
Monee, IL
03 November 2026

67385847R00125